International Conference on Electron Devices and Applications (ICEDA 2021)

AA007558

Journal of Physics: Conference Series Volume 2065

Nanjing, China
14 – 16 August 2021

ISBN: 978-1-7138-5036-6
ISSN: 1742-6588

Printed from e-media with permission by:

Curran Associates, Inc.
57 Morehouse Lane
Red Hook, NY 12571

Some format issues inherent in the e-media version may also appear in this print version.

Institute of Physics
Dirac House, Temple Back
Bristol BS1 6BE UK

Phone: 44 1 17 929 7481
Fax: 44 1 17 920 0979

techtracking@iop.org

Additional copies of this publication are available from:

Curran Associates, Inc.
57 Morehouse Lane
Red Hook, NY 12571 USA
Phone: 845-758-0400
Fax: 845-758-2633
Email: curran@proceedings.com
Web: www.proceedings.com

TABLE OF CONTENTS

Preface

2021 International Conference on Electron Devices and Applications (ICEDA 2021) was held online during August 14-16, 2021. Due to the evolving public health concerns and travel advisories issued for the avoidance of non-essential travel in view of COVID-19, the conference ICEDA 2021 has been held virtually with about 37 guests hosted on Zoom video conference platform. It was a fully synchronized online conference that had live Q&A, participants jointly listened to presentations, and attended other live events associated with the conference.

The conference program was organized in 5 sessions, with 7 plenary and 37 specialized papers presented. Scientists from China, Russia, Thailand took part in the scientific forum. Each plenary speech lasted for 30 minutes, including 5 minutes for Q&A while for other sessions, there was 15 minutes for each presentation, including 2 minutes for Q&A. The interaction session was real experience to the participants and they had good exposure with foreign researchers, students and other delegates.

The objective of ICEDA was to provide a platform for engineers, scientists and industrial partners to discuss and disseminate recent research findings in wide aspects of theoretical, experimental, and practical electron devices and applications fields. Professionals, experts, and scholars from various engineering disciplines were invited to share experiences, innovations, achievements and knowledge. The conference brought together academics, researchers, and practitioners with the aim of bridging the gap between theory and practice in all electron devices aspects. All papers in this proceeding were subject to peer-review by conference committee members and international reviewers. The papers were selected for the proceedings based on quality and relevance to the conference.

We express our gratitude to the Organising Committee for its valuable contribution to this successful edition of ICEDA 2021. We also express our sincere gratitude to all people who contributed to the review process and did all their best to enhance the scientific merit and quality of the Proceedings. We are also grateful to the authors for their contributions, to the speakers and participants of the ICEDA 2021. We wish to extend our warmest welcome to all interested participants to the next ICEDA 2022.

Prof. Wei Lei
Southeast University, China

ICEDA Committees

Advisory Committee Chair

Prof. Daohua Zhang (FInstP, FIET), Nanyang Technological University, Singapore

Conference Chair

Prof. Juin J. Liou (FIEEE), Shenzhen University, China

Conference Co-Chair

Assoc. Prof. Jun Wu, Southeast University, China

Program Chair

Prof. Xiaoqing Wen (FIEEE), Kyushu Institute of Technology, Japan

Program Co-Chairs

Prof. Shumin Xiao, Harbin Institute of Technology, Shenzhen, China
Prof. Letian Huang, University of Electronic Science and Technology of China, China
Dr. Jianying Wei, Anlogic, China

Organizing Committee Chair

Prof. Wei Lei, Southeast University, China

Organizing Committee

Qasim Khan, Shenzhen University, China
Abida Perveen, Southeast University, China
Sajid Hussain, Southeast University, China

Publication Chair

Hong Guo, Dalian University of Technology, China

Publicity Committee

Chenguang Zhou, Peking University, China

International Technical Committees

Prof. Tayeb Mohammed-Brahim, University of Rennes 1, France
Prof. Byung Seong Bae, Hoseo University, Korea
Prof. Razika Zair Tala-Ighil, University M'hamed Bougara Boumerdes, Algeria
Prof. Alexander M. Gaskov, Moscow State University, Russia
Prof. Ridvan Durak, Ataturk University, Turkey
Prof. Poi Sim Khiew, The University of Nottingham, Malaysia
Prof. Sotirios Baskoutas, University of Patras, Greece
Prof. Ajay Kumar, Manipal University Jaipur, India
Prof. Duu Sheng Ong , Multimedia University, Malaysia
Prof. Jingwei Zhao, Taiyuan University of Technology, China
Prof. Jaroslaw Jakubowicz, Poznan University of Technology, Poland
Prof. Ramesh Singh, University of Malaya, Malaysia
Prof. R. S. Jadoun, G. B. Pant University of Agriculture and Technology, India
Prof. Rafiq Noorani, Loyola Marymount University, USA
Assoc. Prof. Sha Shiong Ng, Universiti Sains Malaysia, Malaysia
Assoc. Prof.Rusdi Nur, Politeknik Negeri Ujung Pandang, Indonesia
Dr. Hieu Pham Trung Nguyen, University Heights, USA
Dr. Sujun Guan, Tokyo University of Science, Japan
Dr. Suraya Kamil, Universiti Teknologi MARA, Malaysia

Dr. Mas Ayu Hassan, Universiti Malaysia Pahang, Malaysia
Dr. Shin Yee Khoo, University of Malaya, Malaysia
Dr. Ari Handono Ramelan, Sebelas Maret University, Indonesia
Dr. Rahul Davis, Sam Higginbottom University of Agriculture, Technology & Sciences (SHUATS), India
Dr. Ramadhansyah Putra Jaya, Universiti Teknologi Malaysia, Malaysia
Dr. Yang Xi, II-VI Corp., USA

Peer review declaration

All papers published in this volume of Journal of Physics: Conference Series have been peer reviewed through processes administered by the Editors. Reviews were conducted by expert referees to the professional and scientific standards expected of a proceedings journal published by IOP Publishing.

- **Type of peer review: Double-blind**
  - **Conference submission management system:** Easychair Submission System
  - **Number of submissions received: 38**
  - **Number of submissions sent for review: 38**
  - **Number of submissions accepted: 23**
  - **Acceptance Rate (Number of Submissions Accepted / Number of Submissions Received X 100): 57%**
  - **Average number of reviews per paper: 2**
  - **Total number of reviewers involved: 17**
  - **Any additional info on review process:** All received submissions were subjected to double-blind peer reviews, who are expert or have been experiencing in the related field for years. The accepted papers must be revised according to the referees' comments and suggestions, before inclusion in the conference proceedings. Manuscripts submitted should be original & unpublished and written in accordance with the standard IOP paper template.
  - **Contact person for queries:**
    Ms. Ariel Xie (icedaconf@vip.163.com)

Compliance Current-dependent Dual-functional Unipolar and Threshold Resistive Switching in Silver Nanowires-egg Albumen Composites-based Device

Z J Weng[1], Z W Zhao[1,*], H L Jiang[2], and Y Fang[1]

[1] School of Electronic Science and Engineering, Southeast University, Nanjing 210096, P. R. China
[2] State Key Laboratory of Lake Science and Environment, Nanjing Institute of Geography and Limnology, Chinese Academy of Sciences, Nanjing, 210008, P. R. China

Zhao_zw@seu.edu.cn

Abstract. The continued exploration of novel synthetic memristive materials with multifunctional properties is critical for future synapse-emulating circuits and electronic devices in the field of next-generation neuromorphic computing applications. In this work, the silver nanowires (AgNWs)-Egg albumen composites have been integrated as a resistive switching layer in the Ag/AgNWs-Egg albumen/Ag planar structure and exhibits both unipolar (memory) switching and threshold switching functions. The device in unipolar switching regime demonstrates an ON/OFF ratio above 10^5, a low resistance state of about 1.2 KΩ and a high resistance state of about 120 MΩ. Finally, a mechanism in combination with the conductive filament theory and a tunnelling conduction mechanism is proposed to explain the resistive switching behavior. The devices are prepared by simple and low-cost techniques, which make such devices appealing for future electronic applications.

1. Introduction

With the enormous growth in data and internet of things, the next-generation memory devices able to be scaled beyond the Moore's law are developing more rapidly where memory and processing are better collocated.[1, 2] The emerging memristor which presents a variable resistance that can be controlled by applying a voltage has become a dominant candidate for information storage components and brain-inspired computing.[3-6] In the memristor, the active layer usually changes its states from a high resistance state (HRS or OFF-state) to a low resistance state (LRS or ON-state) and vice versa upon application of voltage amplitudes. Depending on the polarity of the voltage required, the switching process is mainly categorized as bipolar (BRS) and unipolar switching (URS), corresponding to nonvolatile memory switching (MS). Besides the aforesaid MS, there exists another type of switching, threshold switching (TS), which exhibits volatile resistive switching from HRS to LRS at a particular voltage and switches to HRS once the applied voltage is smaller than a critical value.[7, 8]

The use of composite materials in memristor devices has the potential to outperform conventional materials. Their fascinating properties are also favorable for the development of future multifunctional memristor devices. Silver nanowires (AgNWs) have attracted considerable interest in the use of composite materials for memristor technology, where the switching action is controlled by the properties of both the electrodes using AgNWs as core and the dielectric materials coating on the surface of AgNWs that comprise these devices.[9-15] As a typical natural biomaterial, egg albumen (or egg white) has shown promise as the resistive switching layer in the memristors with sandwiched structure.[16, 17] Meanwhile, from the perspective of biological applicability, egg albumen is also biocompatible, implantable, and cheaply produced. Silver nanowire and Egg albumen have already been widely explored, respectively, but to date, the electrical behaviors of AgNWs-Egg albumen composites have not been described.

In this work, we describe the unique electrical properties that emerge by combining the two materials into composites for the first time. Typical memristor cells are constructed as Ag/AgNWs-Egg albumen/Ag planar structure and have displayed a wealth of resistive switching behaviors including the coexistence of both URS and TS, which is distinct from the normal resistive switching behavior of memristor device using egg albumen as the resistive layer. The dual-mode switching behaviors in the devices were thoroughly investigated by controlling the compliance current of the device. On the basis of the results, we propose a switching mechanism in a combined tunneling conduction mechanism and the filament theory to explain the resistive switching behaviors.

2. Experimental Section

2.1. Materials

Silver nitrate ($AgNO_3$, weight-average molecular weight (Mw) = 169.87) was purchased from Shanghai Shenbo Chemical Co., Ltd. Polyvinyl pyrrolidone (PVP, Mw = 1 300 000) was purchased from Tianjin Heowns Biochemical Technology Co., Ltd. Sodium chloride (NaCl, Mw = 58.4) and ethylene glycol (EG, ρ = 1.036 g/mL) were purchased from Chengdu Kelong Chemical Co., Ltd. Egg albumen was obtained from a fresh chicken egg. All chemicals except eggs were used without further purification. Ultrapure, deionized water was used in all experiments.

2.2. Fabrication of the resistive switching device based on AgNWs-Egg albumen composites

The AgNWs were synthesized by a salt-mediated polyol process. In a typical synthesis, ethylene glycol containing PVP was first heated with magnetic stirring at 160 °C for 30 min and then NaCl was added. After cooling to room temperature, $AgNO_3$ was added and then heated at 150 °C for 5 h. Finally, the as-prepared AgNWs were obtained by centrifugation and washed with water and ethanol three times before finally being dispersed in deionized water. Egg albumen was collected by breaking the shell of a fresh, precleaned chicken egg. Some egg albumen can be filtered out by a stainless-steel mesh spoon. Then, extract the sample (volume: 1 mL) from the egg albumen solution using a pipette and mixed with 15 mL of AgNWs solution to synthesize a thin film for device fabrication. The mixed solution was stirred for 30 min. The as-prepared AgNWs-Egg albumen solution was spin-coated onto the precleaned glass substrates with prepatterned Ag electrodes at 1500 rpm for 10 s and 2500 rpm for 20 s. The distance between counter electrodes and width of the electrodes were both about 200 μm for the network device. The device was then baked at 105 °C for 10 min.

2.3. Materials and device characterization

Scanning electron microscopy (SEM) images were taken on a FEI Quanta 200 scanning electron microscope. Transmission electron microscopy (TEM) images and high-resolution transmission electron microscopy (HRTEM) images were observed with Titan 80-300 transmission electron microscope. The crystal structures of the as-synthesized AgNWs were analysed using X-ray diffraction (XRD) with Cu-kα radiation. The electrical properties of the devices based on the AgNWs-Egg albumen composites was tested by a Keithley 4200 Semiconductor Characterization System connecting with a probe station.

3. Results and discussion

SEM and TEM micrograph of the AgNWs is presented in figure 1a and b, respectively. The as-synthesized AgNWs had an average length of about 20 μm and an average diameter of 80 nm. A 3 nm thick PVP coating can be observed in the HRTEM image shown in figure 1c. The AgNWs exhibit a pure face-centered-cubic (fcc) structure (JCPDS File No. 04-0783), as shown in figure 1d.[11]

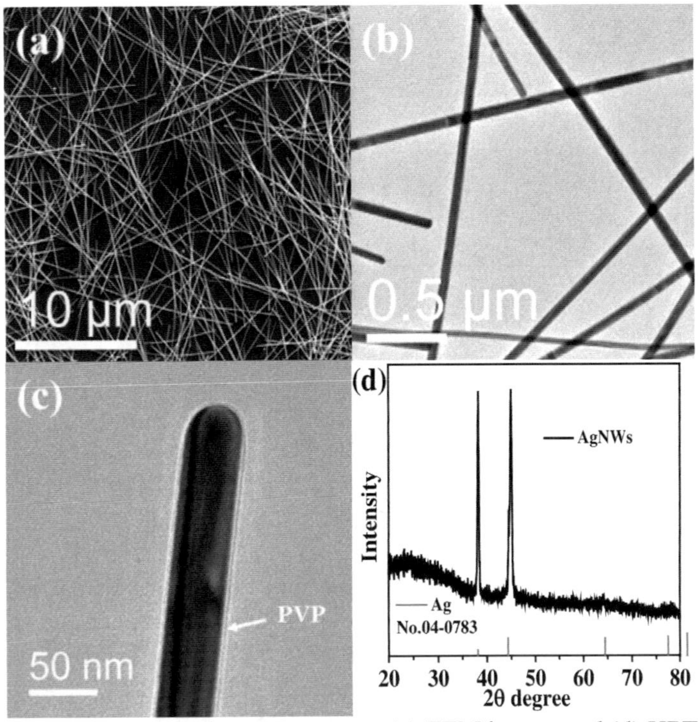

Figure 1. (a) SEM image, (b) XRD pattern, (c) TEM image, and (d) HRTEM image of AgNWs.

Figure 2a shows the current-voltage (I-V) of AgNWs-Egg albumen composites-based device with planar Ag/AgNWs-Egg albumen composites/Ag structure under repetitive positive and negative voltage sweeps with a compliance current (I_{CC}) of 1 μA in linear scale. For the first voltage sweep under positive region, the device starts transition from HRS to LRS at a threshold voltage, V_{th} = +1.16 V, and again switches to HRS for voltage less than a certain voltage, known as the hold voltage, V_{hold} = +0.06 V. The observed behavior corresponds to TS behavior, as the LRS can hardly be maintained after removal of the applied voltage.[18, 19] Symmetric resistive switching behavior was also observed with the opposite polarity bias, showing that the threshold switching is bidirectional in nature. Figure 2b shows the multilevel I-V curves of the device under different I_{CC} values, ranging from 10 to 500 μA. It is worth noting that volatile TS behavior cannot be changed here by a higher applied I_{CC} until increased up to 500 μA. After a higher compliance current (500 μA), during the positive voltage sweep, the current gradually increased, and then, abrupt change of the current was observed at about V_{th} = 0.64 V. However, when the voltage was swept back to 0 V, the device was remained in the LRS. In the negative bias region, the device with switchable resistance was then changed into a resistor, and the devices became conductive. When the compliance current was removed, at a voltage, V_{RESET} = 3.88 V, the device revert to HRS as shown in figure 2c, consistent with RESET process of unipolar resistive switching operation.

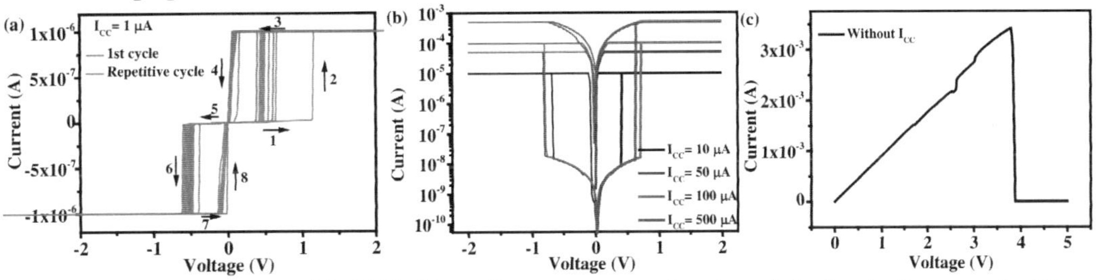

Figure 2. (a) I-V curve of AgNWs-Egg albumen composites-based device in linear scale. The compliance current (I_{CC}) is 1 μA. (b) Multilevel I-V curves of the device operated for four I_{CC} values (I_{CC} = 10, 50, 100, 500 μA) in log scale. (c) I-V curves with I_{CC} being removed for RESET.

Figure 3a shows the typical MS property of the AgNWs-Egg albumen composites-based device in the unipolar regime. During the forming and SET processes, a I_{CC} level of 500 μA is set to prevent the device from switching back to the HRS at zero bias. The device switches from the HRS (~ 120 MΩ) to LRS state (~ 1.2 KΩ) at a SET voltage of V_{set} = +1.88 V. A RESET process was required to bring the device from the LRS to HRS. Thus, I_{CC} is removed, and the RESET process happened at 1.1 V. Figure 3b depicts the data retention capability of the device up to 550 s at a read voltage of 0.5 V, which signifies the nonvolatile property of the memory device. Although little variation in OFF and ON-state can be observed, a steady ON/OFF ratio of 10^5 is maintained in the switching device.

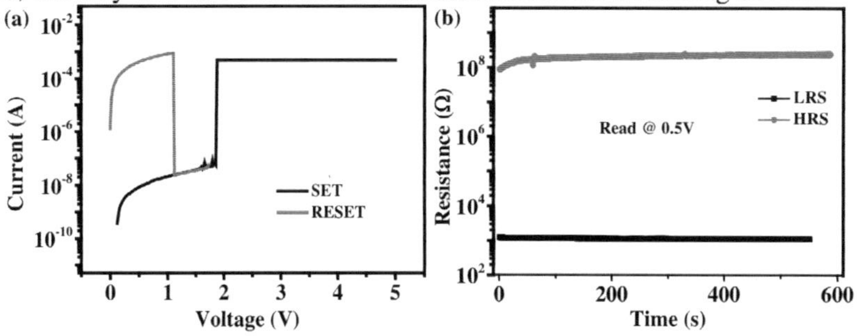

Figure 3. (a) MS property of the AgNWs-Egg albumen composites-based device in the unipolar regime. (b) Retention characteristics of the device.

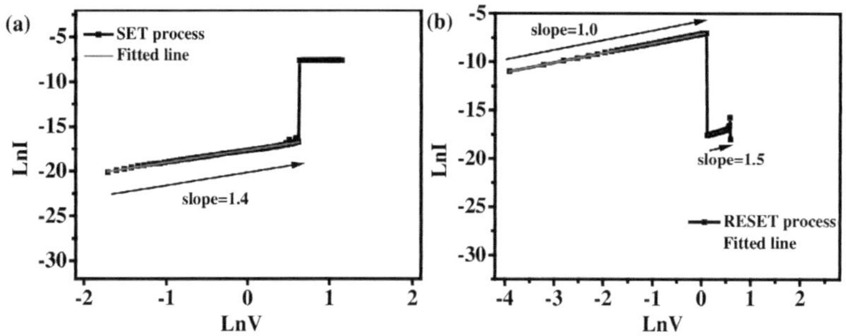

Figure 4. Double logarithmic I−V curves of the AgNWs-Egg albumen composites-based device during the (a) SET and (b) RESET process.

To understand the conduction mechanism of the AgNWs-Egg albumen composites-based device, the I−V characteristic was replotted on the Ln(I) vs Ln(V) graph, as shown in figure 4. During the SET process (Figure 4a), the I−V relationship with a slope of ~1.4 in HRS was observed. After the abrupt increase in the current, switching from HRS to LRS occurred. As shown in figure 4b, LRS showed a nearly linear dependence of current on voltage (~1.0), indicating the ohmic conduction. An obvious RESET process occurred at a lower voltage, and the device was completely switched back to HRS with a slope of ~1.5. It is very common to see two or even three different mechanisms dominate the resistive switching behavior.[20] As a consequence, we propose a possible conduction mechanism based on two different phenomena: i) electron tunnelling over the physical distance of extremely close silver nanowires; and ii) conductive filament formed in the egg albumen. As the applied voltage increases, starting from zero bias at the HRS, the transition to the LRS occurs at a certain threshold voltage due to an abrupt electron tunnelling between close silver nanowires. Besides, as the temperature of the system increases due to Joule heating, thermally-activated electrons and the dissolution of the Ag metal cation from the AgNWs can be injected through the egg albumen under the combined effect of electric fields, thermal diffusion. Once the process starts taking place, it induces a further Joule heating, thus developing a chain reaction process that yields a sharp increase in current. This allows a stable conductive state as a result of conductive filament formation.

4. Conclusion

In summary, it is demonstrated that the AgNWs-Egg albumen composites-based devices display compliance current-dependent dual-functional unipolar and threshold resistive switching functions.

The TS appears for a low compliance current, whereas the MS is initiated by setting a higher compliance current. The conduction mechanism was analysed by electron tunnelling conduction and conductive filament conduction. Although further investigation into the physical mechanisms and performance-limiting factors underlying the resistive switching in this class of composites is needed, an even more significant outcome of this work is that it profoundly inspires new devices that capitalize on this responsive behavior in AgNWs-Egg albumen composites. These results also provide a foundation to better control of the two resistive switching behaviors to establish high-performance emerging devices.

Acknowledgments
This work was financially supported by the National Natural Science Foundation of China (51879042, 51839011); Postgraduate Research & Practice Innovation Program of Jiangsu Province, the Fundamental Research Funds for the Central Universities (KYCX20_0105); State Key Laboratory of Nuclear Power Safety Monitoring Technology and Equipment (K-A2020.415); National Key Research and Development Program of China (2018YFE0125500).

5. References

[1] Sebastian A, Le Gallo M, Khaddam-Aljameh R and Eleftheriou E 2020 *Nat. Nanotechnol.* **15** 529-44
[2] Liu C, Chen H, Wang S, Liu Q, Jiang Y G, Zhang D W, Liu M and Zhou P 2020 *Nat. Nanotechnol.* **15** 545-57
[3] Lin P, Li C, Wang Z, Li Y, Jiang H, Song W, Rao M, Zhuo Y, Upadhyay N K, Barnell M, Wu Q, Yang J J and Xia Q 2020 *Nat. Electron.* **3** 225-32
[4] Zidan M A, Strachan J P and Lu W D 2018 *Nat. Electron.* **1** 22-9
[5] Xia Q and Yang J J 2019 *Nat. Mater.* **18** 309-23
[6] Sangwan V K and Hersam M C 2020 *Nat. Nanotechnol.* **15** 517-28
[7] Sun H, Liu Q, Li C, Long S, Lv H, Bi C, Huo Z, Li L and Liu M 2014 *Adv. Funct. Mater.* **24** 5679-86
[8] He L, Liao Z M, Wu H C, Tian X X, Xu D S, Cross G L, Duesberg G S, Shvets I V and Yu D P 2011 *Nano. Lett.* **11** 4601-6
[9] Manning H G, Niosi F, da Rocha C G, Bellew A T, O'Callaghan C, Biswas S, Flowers P F, Wiley B J, Holmes J D, Ferreira M S and Boland J J 2018 *Nat. Commun.* **9** 3219
[10] Diaz-Alvarez A, Higuchi R, Sanz-Leon P, Marcus I, Shingaya Y, Stieg A Z, Gimzewski J K, Kuncic Z and Nakayama T 2019 *Sci. Rep.* **9** 14920
[11] Du H, Wan T, Qu B, Cao F, Lin Q, Chen N, Lin X and Chu D 2017 *ACS Appl. Mater. Interfaces.* **9** 20762-70
[12] Kim Y, Jeon W, Kim M, Park J H, Hwang C S and Lee S-S 2020 *Appl. Mater. Today.* **19** 100569
[13] Manning H G, Biswas S, Holmes J D and Boland J J 2017 *ACS Appl. Mater. Interfaces.* **9** 38959-66
[14] Kuncic Z and Nakayama T 2021 *Adv. in Phys.: X.* **6** 1894234
[15] Li Q, Diaz-Alvarez A, Iguchi R, Hochstetter J, Loeffler A, Zhu R, Shingaya Y, Kuncic Z, Uchida K i and Nakayama T 2020 *Adv. Funct. Mater.* **30** 2003679
[16] Zhou G, Yao Y, Lu Z, Yang X, Han J, Wang G, Rao X, Li P, Liu Q and Song Q 2017 *Nanotechnology* **28** 425202
[17] Yan X, Li X, Zhou Z, Zhao J, Wang H, Wang J, Zhang L, Ren D, Zhang X, Chen J, Lu C, Zhou P and Liu Q 2019 *ACS Appl. Mater. Interfaces.* **11** 18654-61
[18] Wan T, Pan Y, Du H, Qu B, Yi J and Chu D 2018 *ACS Appl. Mater. Interfaces.* **10** 2716-24
[19] Wang Z, Rao M, Midya R, Joshi S, Jiang H, Lin P, Song W, Asapu S, Zhuo Y, Li C, Wu H, Xia Q and Yang J J 2017 *Adv. Funct. Mater.* **28** 1704862
[20] Zhu J, Zhang T, Yang Y and Huang R 2020 *Appl. Phys. Rev.* **7** 011312

The Quantum Calculation for Valence Band Structure of Strained Zinc-blende GaN Using Six-Band Based k·p Method

Yaqun Liu, Everett X. Wang, Gary Zhang and Xiyue Li*

School of Information Engineering, Guangdong University of Technology, Guangzhou, China

lixiyue915@gdut.edu.cn

Abstract. The variations of valence band energy with stress effects in zinc-blende GaN are proposed in this paper. The calculations are based on a six-band strain dependent k·p Hamiltonian, and can be self-consistently solved by Schrödinger-Poisson equation. Accurate physical pictures are given for the quantized valence subband structure under biaxial and uniaxial stress in (001) surface along the [110] direction accounting the quantum confinement effect. The warping of the energy profile results in carrier distribution change. This research will be beneficial for improving the hole mobility and the selective of optimum stress for group-III nitride semiconductor based devices.

1. Introduction

As one of the representatives of the Wide-Bandgap semiconductor materials, GaN have attracted significant amounts of attention in the past several decades due to the promising electrical and optical properties [1-3]. It is known that the polarization charge in zinc-blende GaN is less than wurtzite form, and the studies focus on valence band structure is also relatively few. However, as GaN play a crucial role in 5G devices and LEDs, and the zinc-blende structure is superior to the wurtzite structure in terms of symmetry, ease of doping, and low phone scattering. Therefore, it is still very important to study the potential of zinc-blende form, especially for the properties in heterostructure taking the stress effect which is caused by AlGaN substrate into account.

In the earlier, the band structure of GaN has been calculated by using methods such as first-principles calculation [4], the empirical pseudopotential method (EPM) [5], quasiparticle calculation [6] and so on. In this work, we use the six-band stress-dependent quantum k·p method to calculate the valence band structure of zinc-blende GaN. With this method, both biaxial and uniaxial effects are investigated. Quantum confinement effect is also included. The energy band structure is the foundation in fully appreciating the properties of materials, and the dispersion relation of the valence band in strained materials is also the theoretical basis for understanding the valence band properties and calculating hole mobility. This analysis will provide a physical explanation for the variation of valence band structure under different stress conditions.

2. Calculation for valence band structure

The k·p method provides a powerful approach for calculating valence band structure in stressed zinc-blende GaN. According to Luttinger-Kohn theory, the valence band structure for Bulk zinc-blende GaN can be obtained by the following 6×6 Hamiltonian matrix [7][8]:

$$\begin{vmatrix} Q & S & R & 0 & \frac{i}{\sqrt{2}}S & -i\sqrt{2}R \\ S^* & T & 0 & R & -\frac{i}{\sqrt{2}}(Q-T) & i\sqrt{\frac{3}{2}}S \\ R^* & 0 & T & -S & -i\sqrt{\frac{3}{2}}S^* & -\frac{i}{\sqrt{2}}(Q-T) \\ 0 & R^* & -S^* & Q & -i\sqrt{2}R^* & -\frac{i}{\sqrt{2}}S^* \\ -\frac{i}{\sqrt{2}}S^* & \frac{i}{\sqrt{2}}(Q-T) & i\sqrt{\frac{3}{2}}S & i\sqrt{2}R & \frac{1}{2}(Q+T)-\Delta_{so} & 0 \\ i\sqrt{2}R^* & -i\sqrt{\frac{3}{2}}S^* & \frac{i}{\sqrt{2}}(Q-T) & \frac{i}{\sqrt{2}}S & 0 & \frac{1}{2}(Q+T)-\Delta_{so} \end{vmatrix} \quad (1)$$

where

$$\begin{aligned} & Q = Q_k + Q_\varepsilon \quad T = T_k + T_\varepsilon \\ & R = R_k + R_\varepsilon \quad S = S_k + S_\varepsilon \\ & Q_k = -\frac{\hbar^2}{2m_0}[(\gamma_1+\gamma_2)(k_x^2+k_y^2)+(\gamma_1-2\gamma_2)k_z^2] \\ & T_k = -\frac{\hbar^2}{2m_0}[((\gamma_1-\gamma_2)(k_x^2+k_y^2)+(\gamma_1+2\gamma_2)k_z^2] \\ & S_k = i\frac{\hbar^2}{m_0}\sqrt{3}\gamma_3(k_x-ik_y)k_z \\ & R_k = -\frac{\hbar^2}{2m_0}\sqrt{3}[\gamma_2(k_x^2-k_y^2)-2i\gamma_3k_xk_y] \\ & Q_\varepsilon = a_v(\varepsilon_{xx}+\varepsilon_{yy}+\varepsilon_{zz})+0.5b(\varepsilon_{xx}+\varepsilon_{yy}-2\varepsilon_{zz}) \\ & T_\varepsilon = a_v(\varepsilon_{xx}+\varepsilon_{yy}+\varepsilon_{zz})-0.5b(\varepsilon_{xx}+\varepsilon_{yy}-2\varepsilon_{zz}) \\ & R_\varepsilon = \frac{\sqrt{3}}{2}b(\varepsilon_{xx}-\varepsilon_{yy})-id\varepsilon_{xy} \\ & S_\varepsilon = -d(\varepsilon_{yz}-i\varepsilon_{xz}) \end{aligned} \quad (2)$$

m_0 denotes to free electron mass, γ_1, γ_2 and γ_3 are Luttinger parameters proposed in [7], Δ_{so} is the spin-orbit interaction, a_v, b and d are Bir-Pikus deformation potential. In this work, we restrict ourselves to the special case of a biaxial stress and a uniaxial tress along x-y plane, namely,

$$\begin{aligned} & \varepsilon_{xx} = \varepsilon_{yy} \neq \varepsilon_{zz} \\ & \varepsilon_{xy} \neq 0 \\ & \varepsilon_{yz} = \varepsilon_{xz} = 0 \end{aligned} \quad (3)$$

thus

$$S_\varepsilon = 0$$

Therefore, the relationship between strain e_{ij} and stress s_{ij} can be written as

$$\begin{aligned} & \varepsilon_{xx} = s_{11}\cdot s_{xx} + s_{12}\cdot(s_{yy}+s_{zz}) \\ & \varepsilon_{yy} = s_{11}\cdot s_{yy} + s_{12}\cdot(s_{xx}+s_{zz}) \\ & \varepsilon_{zz} = s_{11}\cdot s_{zz} + s_{12}\cdot(s_{xx}+s_{yy}) \\ & \varepsilon_{xy} = s_{44}\cdot s_{xy} \end{aligned} \quad (4)$$

and s_{11}, s_{12}, s_{44} can be related to the elastic constants [9] with the description of

$$\begin{aligned} & s_{11} = [1/(c_{11}+2\cdot c_{12})+2/(c_{11}-c_{12})]/3 \\ & s_{12} = [1/(c_{11}+2\cdot c_{12})-1/(c_{11}-c_{12})]/3 \\ & s_{44} = 1/c_{44} \end{aligned} \quad (5)$$

In this work, the valence band along (001) is classified by angular momentum due to the fourfold rotation symmetry in zinc blende crystals. Solving for the eigenvalue of the Hamiltonian matrix including the spin-orbit interaction, the basis of Bloch states consist of six angular momentum Eigen states, which representing heavy-hole, light-hole and spin-orbit split-off valence band state with spin up and down, can be obtained. However, if the quantization is not included, only 6 subbands can be observed. It is not enough for the researches in electrical and optical properties.

In order to accurately realize the subband structure for zinc-blende GaN, an investigation has been undertaken by using the k·p model approach, which is simple enough to be solved by Schrödinger equations and rigorous enough to include the effect of arbitrary stress. This method has also been used in strained bulk and quantum well structures with zinc-blende symmetry [10].

In GaN/AlGaN heterojunction interface, the k_z-direction is quantized because of the quantum well produced by the gate filed in a (001) wafer. By replacing k_z with an operator $(\hbar / i)(\partial/\partial z)$, the quantized valence hole subband energy can be calculated by using (6):

$$H(k_{x,y},k_z)+V(z)\cdot\varphi_{k_{x,y}}(z)=E(k_{x,y})\cdot\varphi_{k_{x,y}}(z) \qquad (6)$$

where $H(k_{x,y},k_z)$ is the quantized k·p Hamiltonian matrix including the strain component. The (6) yield negative eigenvalues E(k), and below the sign convention appropriate to analyze transport [positive kinetic energies E(k)] will be employed.

Equation (6) can be discretized on a mesh in the z-direction with N_z nodes and turn into an eigenvalue problem of a $6N_z \times 6N_z$ complex band Hermitian matrix to propose the iso-energy contours. With the use of non-uniform mesh which can reduce the computation time effectively, a matrix with 12 non-zero length tridiagonal block form can be obtained and the calculation in this work includes the lowest 30 subbands. As the relative fine discretization employed, the subband structure requires the solution of eigenvalue problem of Schrödinger equations and its discretized form to obtain the quantized subband energies. A self-consistent, two-dimensional solution of the Schrödinger and Poisson equations is utilized in order to analyze a much wider range of account for band profile as rigorously as possible. This method is very efficient in finding Eigen-states that extend over a relatively large spatial area without loss of accuracy.

3. Result and discussion

3.1 Bulk Zinc-blende GaN Valence Band Structure

In this section, we will describe the main effect on the valence band of bulk zinc-blende GaN. The (001) is chosen as the transport plane and [110] as the current direction since it is currently the case for the similar PMOS technology. Two kinds of stress are considered, including the biaxial stress due to lattice mismatch and uniaxial stress along the current channel.

In Fig. 1(a) and (b), we can see the three-dimensional heavy-hole (HH) iso-energy structure near Γ point without applied stress. The HH band is consist of 12 "wings" including four "in-plane" wings located in the plane of the channel and eight "out-of-plane" wings understood as perpendicular to the channel. In no-stress case, all wings are equivalent at energy, and also be equally populated by hole carriers. The calculated HH band with biaxial tensile, uniaxial tensile and uniaxial compressive are shown in Fig .1(c) ~ (e), describing that the energies are not equivalent at all. The effects causing by biaxial and uniaxial stress will be discussed later.

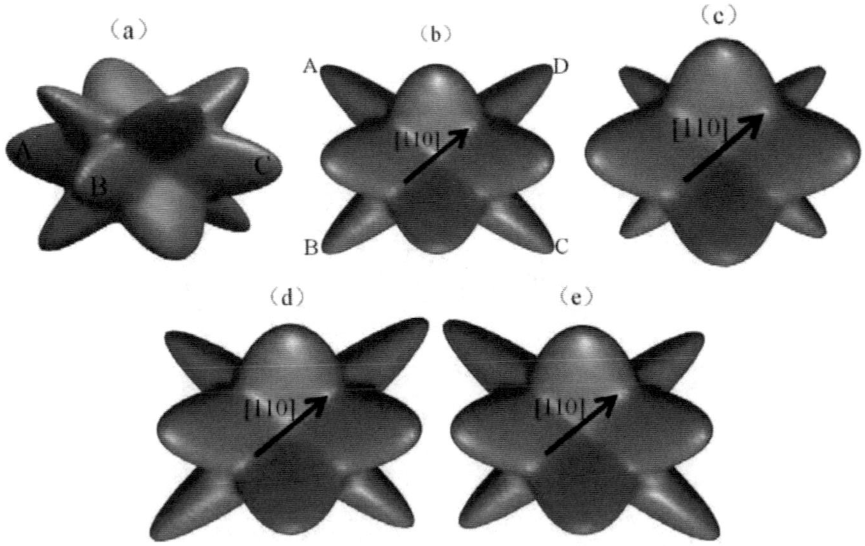

Figure 1. The bulk zinc-blende GaN HH iso-surfaces at 0.5eV under (a) (b) no stress, (c) biaxial tensile, (d) uniaxial tensile, and (e) uniaxial compressive. The thick arrow indicates the current direction. (b), (c), (d) and (e) are from top views. The arrow indicates the current direction [110].

3.2 Quantized Subband Structure

To further illustrate the quantum well structure including the effects of biaxial and uniaxial stress, the discretized E-k dispersion of zinc-blende GaN under a gate field are calculated, taking quantum confinement effect into account. By solving (6), the quantized subband energy level can be obtained. Fig. 2(a) ~ (f) shows iso-energy contours of the lowest 6 quantized subbands. We can see that only the shapes of first and second subbands are similar to Fig.1 (b), and the more the energy, the more the shapes deviates. Therefore, it is feasible to simulate the properties of quantum well by using the first pair subbands.

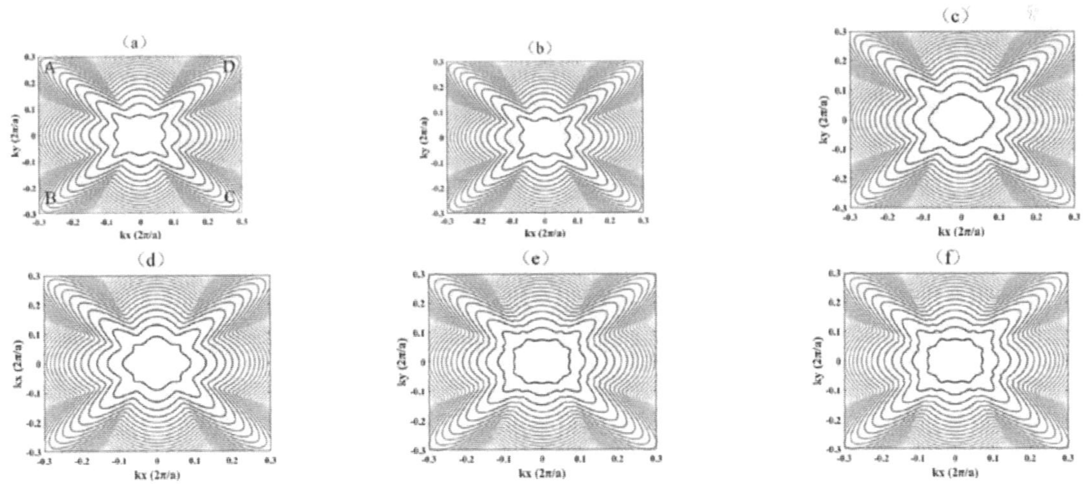

Figure 2.Iso-energy contours of the lowest six quantized subbands separated by 25meV under no stress condition

From Fig. 2(a), we can observe the strong anisotropy of iso-surface. In $k_z \neq 0$ case, all the shapes which can be observed in the bulk bands are still visible in the picture, while the band below the plane cannot be seen because it is the same as the band above the plane. When these "out-of-plane" wings are occupied and there is a vertical confinement field in the *z*-direction, effective coupling can be provided.

Fig.3 shows the iso-energy contour of the first subband in k_z=0 plane is markedly four-angle star shape. The reason for choosing this plane is we can only focus on the four "in-plane" wings which play the most important role in transport behavior.

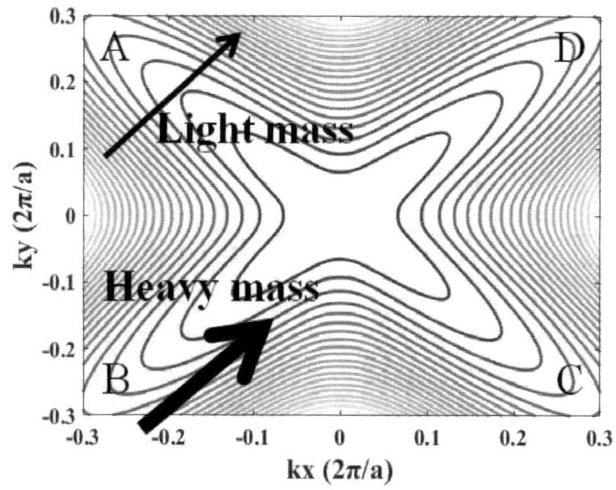

Figure 3.Iso-energy contours of the first subband separated by 25meV in k_z=0 plane without applied stress. It clearly shows high curvature (small effective mass) at A, C, and low curvature (high effective mass) at B, D.

It is clearly that along current channel direction of [110], A and C have more curvature than B and D. Therefore, the effective mass of carriers sitting on A and C is lighter than that sitting on B and D, while the carriers on the out-of-plane wings have medium effective mass. Under different stress, wings energy can be moved up and down. Some wings will become dominant when their energy is lowered by stress. Since all the wings have their minima at Γ point. The dominant wing with the lowest energy has the highest population concentration at and around the Γ point. Therefore, we expect that the band shape of the leading wings also determines the hole distribution and mobility behavior.

Next, we examine the effects induced by biaxial and uniaxial stress. The lattice mismatch between the epitaxially grown GaN and AlGaN substrate will generate biaxial stress. Fig. 4 shows the biaxial tensile (s_{xx}=s_{yy}=1GPa) case (the solid contours) and unstressed case (the dotted contours). An interesting phenomenon is that the innermost solid contours are circular rather than four-angle star shaped. It is revealed that the energy of light hole (LH) band is now below the heavy hole (HH) band under biaxial tensile stress, and the rotation symmetry along the *z* axis is unchanged.

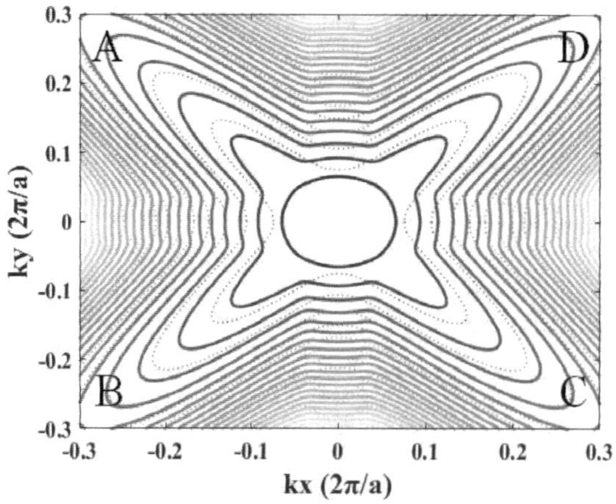

Figure 4.Iso-energy contours of the first subbands separated by 25 meV in k_z=0 plane under 1GPa biaxial tensile stress

The uniaxial stress consists of biaxial and shear part, therefore the model has to be able to treat both biaxial and shear stress to be accurate. When the GaN/AlGaN heterojunction is stressed along the channel direction, the uniaxial stress includes the s_{xx} component along the channel, the s_{yy} component perpendicular to the channel, and s_{xy}, which is a shear. The presence of the s_{xy} component describes the main difference between biaxial and uniaxial stresses, and it can also induce the carrier repopulation among the "in-plane" wings. Fig. 5 and Fig. 6 show the iso-energy contours of the first subbands in (001) plane for lowest energy hole band under 1GPa uniaxial compressive stress (s_{xx}=s_{yy}=s_{xy}=-500MPa) and uniaxial tensile stress (s_{xx}=s_{yy}=s_{xy}=500MPa) in [110] direction, respectively. The dotted contours on each plot also show the non-stress case. Uniaxial stress changes the rotation symmetry about the *z*-direction to twofold. When the 1GPa uniaxial stress is applied, the iso-energy contours of the innermost region are close to an "ellipsoid-like" shape. The distortion of the iso-contour present that energy distribution will change. In particular, Fig. 5 shows an energy lowering in A and C, while a rise can be seen in B and D with uniaxial compressive stress. This results a carrier repopulation from B and D to A and C, causing a mobility enhancement because the lighter effective mass is advantageous to transport characteristics. On the other hand, with the uniaxial tensile stress in Fig.6, the result is opposite, the carrier are redistributed from A and C to B and D, leading the mobility degradation. The same results can be also certified in Fig. 7. Presenting that under the uniaxial compressive stress condition, the energy of the A-C line is much closer to Γ point than B-D line. Therefore, uniaxial compressive stress is considered to be the most efficient way to enhance hole mobility with the combined effects by biaxial compression and shear components.

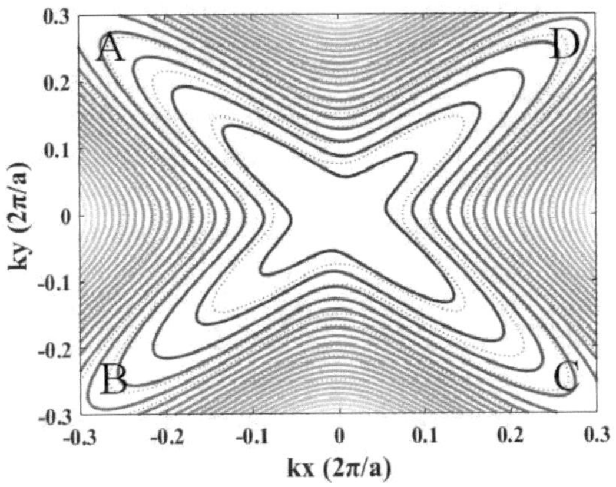

Figure 5.Iso-energy contours of the first subbands separated by 25 meV in k_z=0 plane under 1GPa uniaxial compressive stress

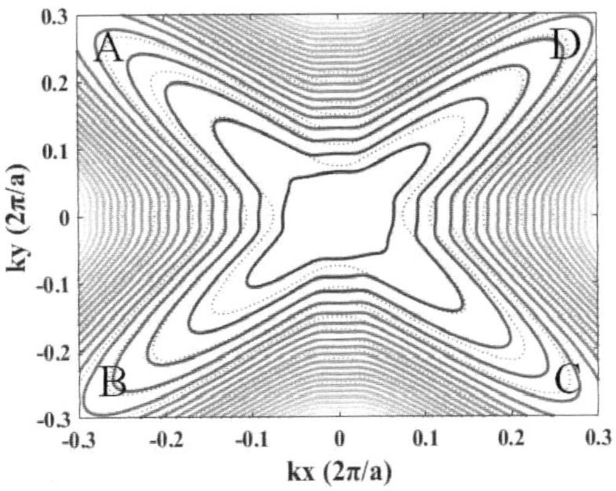

Figure 6.Iso-energy contours of the first subbands separated by 25 meV in k_z=0 plane under 1GPa uniaxial tensile stress

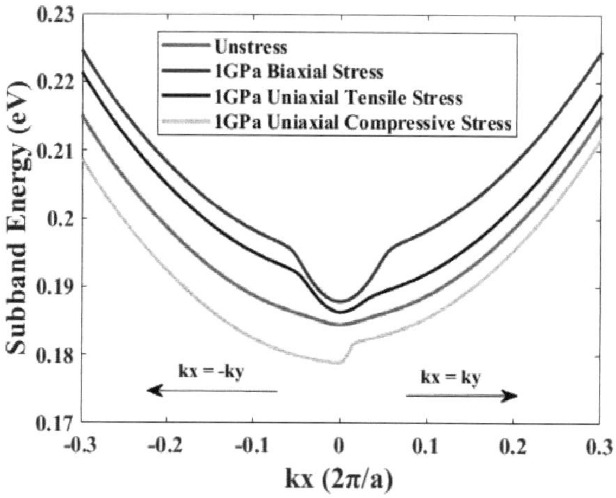

Figure 7. Energies of the first subband in A-C line and B-D line with no stress, 1GPa biaxial tensile stress, 1GPa uniaxial compressive stress and 1GPa uniaxial tensile stress condition, respectively.

4. Conclusion

In this work, a concise overview of the quantized subbands model and applicable for zinc-blende GaN/AlGaN heterojunction quantum well has been presented by employing a six-band based stress-dependent k·p method, and presented their respective work mechanisms. Based on this model, comprehensive analysis of the energy dispersion for the valence band structure under biaxial and uniaxial stress in (001) surface along [110] direction has been given. The variation of subband energies is illustrated by two-dimensional iso-energy diagram. Uniaxial compression in the [110] direction can enhance hole mobility through the combination of biaxial compressive and the shear stress. The biaxial compression component can make the holes redistribute into four "in-plane" wings, and the shear component further move holes into the wings perpendicular to the channel, reducing the effective mass greatly. Such combined effects are considered to make uniaxial compressive stress as the most efficient way for the hole mobility enhancement in zinc-blende GaN/AlGaN heterojunction.

The method in this work will be very useful for the calculation of valence band in strained quantum well structure, which has a great potential for applications to optical and electronic devices.

Acknowledgment

This work is supported by The National Key Research and Development Program of China (2018YFB1802100).

5. References

[1] O. Ambacher, "Growth and applications of group III-nitrides," Journal of Physics D: Applied Physics, vol. 31(20), 1998, pp. 2653-2710, doi: 10.1088/0022-3727/31/20/001.

[2] S. Saib, N. Bouarissa, P. Rodríguez-Hernández, and A. Munoz, "Structural properties of zinc-blende $Ga_xIn_{1-x}N$: ab initio calculations, " European Physical Journal B, vol. 60, 2007, pp: 435-438, doi: 10.1140/epjb/e2008-00015-3.

[3] A. Raj, A. Krishna, N. Hatui, C. Gupta, R. Jang, S. Keller, and U.K. Mishra, "Demon□stration of a GaN/AlGaN superlattice-based p-channel FinFET with high ON current," IEEE Electron Device Letters, vol. 41, 2020, pp. 220-223, doi: 10.1109/LED.2019.2963428.

[4] H. Y. Xiao, X. D. Jiang, G. Duan, Fei Gao, X. T. Zu, and W. J. Weber, "First-principles calculations of pressure-induced phase transformation in AlN and GaN," Computational Materials Science, vol. 48, 2010, pp. 768-772, doi: 10.1016/j.commatsci.2010.03.028.

[5] M. L. Cohen, and T. K. Bergstresser, "Band structures and pseudopotential form factors for fourteen semiconductors of the diamond and zinc-blende structures," Physical Review, vol. 141, 1996, pp. 789-796, doi: 10.1103/PhysRev.141.789.

[6] A. Rubio, J. L. Corkill, M. L. Cohen, E. L. Shirley, and S. G. Louie, "Quasiparticle band structure of AlN and GaN," Physical review B, vol. 48, 2003, doi: 10.1103/PhysRevB.48.11810.

[7] D. Fritsch, H. Schmidt, and M. Grundmann, "Band-structure pseudopotential calculation of zinc-blende and wurtzite AlN, GaN, and InN," Physical review B, vol. 67, 2003, pp. 235205(1-13), doi: 10.1103/PhysRevB.67.235205.

[8] P. Sengupta, H. Ryu, S. Lee, Y. Tan, G. Klimeck, "Numerical guidelines for setting up a general purpose k.p simulator with applications to quantum dot heterostructures and topological insulators, " Journal of Computational Electronics, vol. 15, 2014, doi: 10.1007/s10825-015-0729-6.

[9] A. Wright, "Elastic properties of zinc-blende and wurtzite AlN, GaN, and InN," Journal of Applied Physics, vol. 82, 1997, doi: 10.1063/1.366114.

[10] M. Fischetti, Z. Ren, P. Solomon, M. Yang and K. Rim, "Six-band k·p Calculation of the Hole Mobility in Silicon Inversion Layers: Dependence on Surface-Orientation, Strain, and Silicon Thickness," Journal of Applied Physics, vol. 94, 2003, doi: 10.1063/1.1585120.

The Effect of UV Irradiation on Tensile Strength of Polypropylene Reinforced Kenaf Fiber and Peat Soil

Nurul Farahin Mohd Joharudin[1], Mohammad Sukri Mustapa[1,2,*], Aiman Supawi[1], Noradila Abdul Latif[1,2], Wan Nasrul Nizam Wan Hussin[1], Mohd Fairuz Alias[1], Kamarul-Azhar Kamarudin[1], Ahmad Hamdan Ariffin[1] and Farazila Yusof[3]

[1]Faculty of Mechanical and Manufacturing Engineering, Universiti Tun Hussein Onn Malaysia, 86400 Parit Raja, Batu Pahat, Johor, Malaysia
[2]Mechanical Failure Prevention and Reliability (MPROVE), Universiti Tun Hussein Onn Malaysia, 86400 Parit Raja, Batu Pahat, Johor, Malaysia
[3]Faculty of Engineering, University of Malaya, 50603 Wilayah Persekutuan Kuala Lumpur, Malaysia

sukri@uthm.edu.my

Abstract. Nowadays, natural fiber polymer composites have been widely used in many industries and applications because of their low cost, renewable and eco–friendly. However, exposing this material to the outdoor environment will affect the material properties as it is exposed to various situations such as ultraviolet exposure, raining, etc. This paper studies the degradation of kenaf fiber, polypropylene strengthened with peat soil combustion at 600ºC upon ultraviolet exposure. The wood–plastic composites (WPCs) produced into two batches, kenaf fiber mixing with polypropylene and kenaf fiber and peat soil mixing with polypropylene. All the specimen has exposed in ultraviolet (UV) irradiation at 0 hour, 100 hours, and 200 hours, respectively. The UV irradiation accelerated weathering tester machine was conducting the water spray cycle, which is 4 hours of water spray and 8 hours without water spray continuously until achieved the target time. The finding obtained from the tensile test, the composition with the highest contain of kenaf fiber in a polymer composite, has the lowest tensile strength after exposed with 200 hours of UV irradiation, which is 20.23 MPa. While the adding peat soil as reinforcement has shown the increasing the percentage of tensile strength after 200 UV exposure.

1. Introduction

Wood is made from natural polymers such as lignin, cellulose, and various hemicelluloses, but its properties vary significantly from those of the most commonly mixed synthetic polymers [1]. The effective structure and morphology make it a rigid, solid, durable, and lightweight material that can effectively perform functions that are necessary for the survival of the tree, such as moisture transport. For decades, it has been a useful structural material due to its excellent performance and low cost. However, some of the same material behavior in a composite material, such as moisture sorption, that occurs naturally could be troublesome [2]. To enable the effective use of wood as a filler or reinforcement in polymers, an understanding of its material behavior is essential.

Wood–plastic composites (WPCs) are being used, showing, in reality, an extremely wide variety of possible applications on the market in many sectors [3-5]. In outdoor construction, about two-thirds of

WPCs are used as decking, siding, and roof tiles [6]. Degradation of WPCs can result in outdoor use of abiotic agents such as moisture, sunlight, and temperature [6,7]. Fading color and loss of WPCs mechanical properties caused by outdoor weathering has become a significant concern, and the WPCs weathering accountability mechanisms need to be enhanced [8]. Methods developed to increase their weathering resistance.

In this study, WPCs weather capability was assessed through natural and accelerated weathering monitoring. WPCs discoloration (whitening) was caused by the deterioration of both wood and plastic during test exposure. Darker color pigments as additives enhanced WPCs color stability, but chalking was still observed on the surfaces. WPCs color consistency was improved during the pre–mixing phase by adding the ultraviolet stabilizer. The application of the UV stabilizer in WPCs had a beneficial effect on color consistency and prohibited external use of the composites chalking.

2. Experimental Material and Procedure

2.1. Materials
Polypropylene is used as polymer reinforcement, while kenaf fiber and peat soil is used as a reinforcement material. This project uses a granulator machine to blend kenaf fiber reinforced with polypropylene with peat soil. Peat soil has been crushed and undergoes the burning process at 600ºC.

2.2. Mixing Process
Peat soil, kenaf fiber, and polypropylene are mixed in this mixing process by using Bra-bender roll mixer machine. The speed parameter was set at 190°C and 200 rpm temperature. For stabilization, approximately 30 minutes time needed. Then it is necessary to pour the polypropylene moderately into the machine and let it liquefy. The combination of peat soil and kenaf fiber is then slowly poured and allowed to melt and blend well for around 6 minutes. The composite component must be taken out immediately to prevent overheating.

2.3. Injection Molding
An injection molding machine was used to produce a specimen in dog bone shape. The process that involves forming the specimen is quite sensitive and needs more focus as its materials are kenaf fiber, polypropylene, and peat soil. The specimen type to be produced will be followed for time-keeping of the process cycle. Therefore, it is very important to set the temperature and observe it frequently to avoid any defect that has a bubble, hollow, or wrapped up the specimen.

2.4. Ultraviolet (UV) Irradiation Test
The UV irradiation accelerated weathering Tester, model LUV was used to measures the materials at regulated and elevated temperatures by exposing them to alternating cycles of UV light and moisture. It simulates sunlight's effects with fluorescent ultraviolet lamps and simulates dew and rain according to the ASTM D4329-99 standard by condensing moisture and/or water spray. Types of damage include color changed, gloss loss, chalking, cracking, crazing, hazing, blistering, fragility, strength loss, and oxidation. It was used to irradiate the specimen into ultraviolet light to mimic the real sunlight at predetermined temperatures and hours.

2.5. Tensile Test
The Universal Testing Machine (UTM) is used to determine the specimen's tensile strength. This testing uses SHIMADZU AGS-J Universal Testing Machine that follows ISO 527-5. This test feature is subjected to controlled stress as a sample before failure occurs. Analysis of the tensile stress was done to get where the plastic specimen would be broken. This experiment was carried out with 1 mm/min crosshead speed parameters that were applied with 5 KN of force that surrounded (25°C) of temperature, where the specimen is gripped by an adjustable upper and lower crosshead at the pump. The upper and lower crossheads move moderately, and the specimen is slowly pulled apart. At the breaking point, the measurement was taken.

3. Results and Discussion

3.1. Color Degradation after UV Exposure

Wood Polymer Composite contains the composition of natural fiber theoretically can provide desirable unique strength, modulus at a lower cost [9-11]. The most significant change is when increasing the composition of kenaf fiber in WPCs. This can be seen starting with the combination of 30% KF + 70% PP and become worst with 40% KF + 60 PP after 200 hours of UV exposure, as shown in Table 1. The biggest change in color due to the higher lignin content in kenaf fibers.

Table 1. Discoloration of tensile specimen after UV exposure.

Samples	Hours		
	0	100	200
10% KF + 90% PP	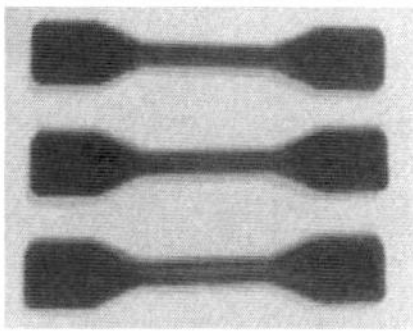	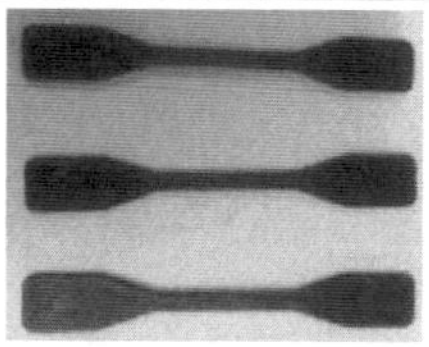	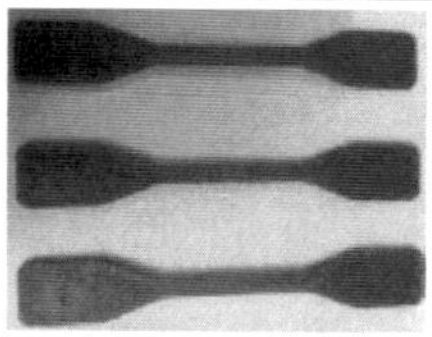
20% KF + 80% PP	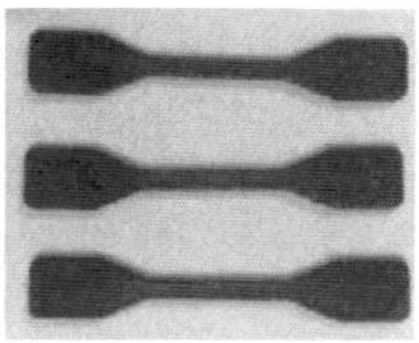	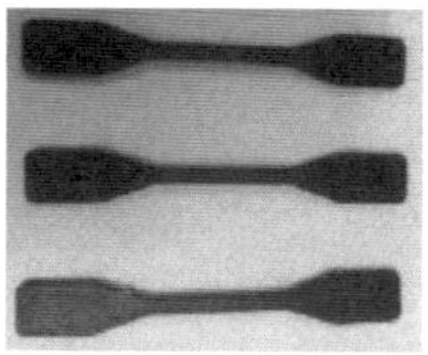	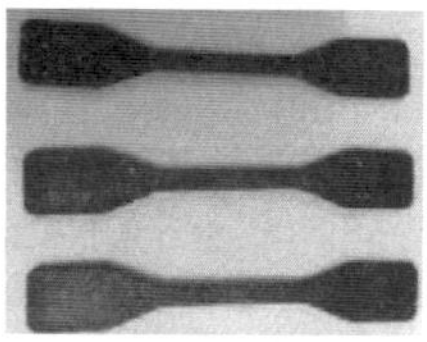
30% KF + 70% PP	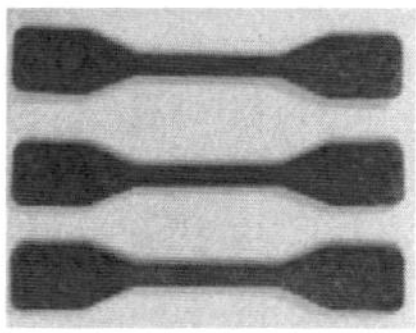	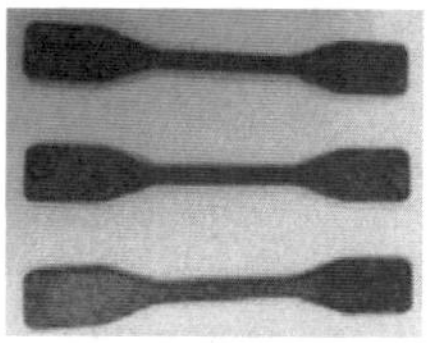	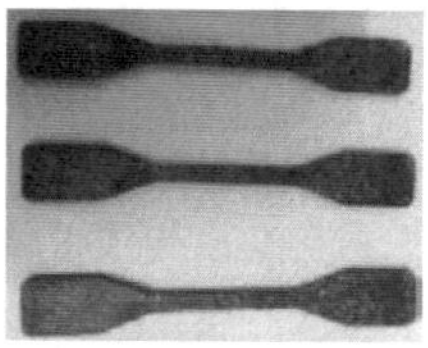
40% KF + 60% PP	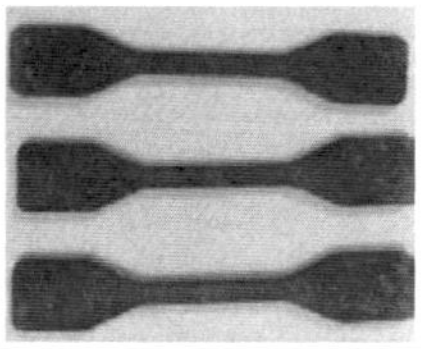	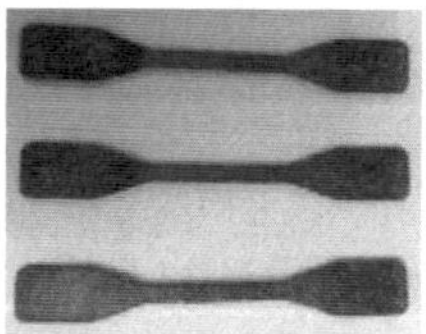	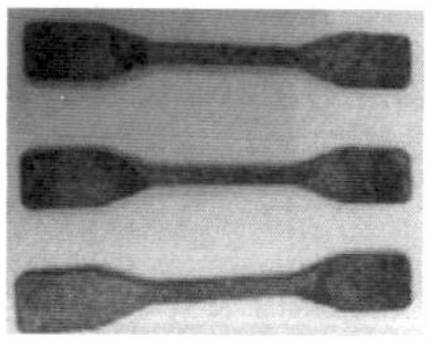

3.2. Tensile Test

The mechanical test has been done for all specimens' carbon peat soil reinforced with kenaf fiber and polypropylene that already done for UV irradiation exposure. This test is be done to investigate which combination of materials still have withstood with the best strength after exposure to UV irradiation.

Table 2 shows the result average of strength of combination kenaf fiber with polypropylene upon UV irradiation exposure. From Figure 1, it can be seen that the graph shows non-linear for all specimens after UV exposure. Before exposed to the UV, the result shows that the best composition for kenaf fiber is 30% mixing with 70% PP with the highest tensile strength, 21.03 MPa. Previous studies have shown that the best composition for kenaf fiber mixing with polypropylene is 30% KF + 70% PP [12]. From observing within this short range of time, which is 0 to 200 hours of UV exposure, some of the specimens changed in terms of mechanical properties. The variation strength and strain changes are because void formation occur between fiber and matrix while exposed with UV irradiation that can affect the flexural strength [13]. The material that shows the increment of strength from 0 to 200 hours of UV exposure is 10% KF + 90% PP, 20% KF + 80 PP, and 40% KF + 60% PP. The only one combination 30% KF + 70% PP experiencing a decrease in strength is about 3.8%, even this composition has the best tensile strength before UV exposure.

After exposure to 200 hours of UV irradiation, the highest maximum strength is 22.78 MPa which is the combination of 10% KF + 90% PP. While the lowest tensile strength after 200 UV exposure is 20.23 MPa, which is 40% KF and 60% PP. This means that the best composition of kenaf fiber with polypropylene within 200 hours of UV exposure is 10% KF + 90% PP. This composition has the least kenaf fiber content will degrade with the increment of UV exposure [14]. This is because in UV accelerated weathering tester, tensile specimen experiencing UV exposure with the cycle of water sprinkle. So once the kenaf fiber composite exposed with the moisture, water penetrates and attaches onto hydrophilic fiber groups that create intermolecular hydrogen bonding with fibers and decreases fiber-to-matrix interfacial adhesion, resulting in poor transfer stress from the matrix to fiber and lower kenaf composite tensile strength.

Table 2. Result average of strength of combination kenaf fiber with polypropylene upon UV irradiation exposure.

Samples	Mechanical Properties	Hour			Decrease/Increase (%)
		0	100	200	
10% KF + 90% PP	Max. strength (MPa)	20.47	22.75	22.78	Increase (11.28)
	Max. strain (%)	10.23	10.14	10.12	Increase (1.08)
20% KF + 80% PP	Max. strength (MPa)	20.43	22.78	21.92	Increase (7.29)
	Max. strain (%)	7.92	7.82	7.02	Decrease (11.36)
30% KF + 70% PP	Max. strength (MPa)	21.03	18.66	20.71	Decrease (1.52)
	Max. strain (%)	5.21	5.13	5.18	Decrease (0.58)
40% KF + 60% PP	Max. strength (MPa)	19.65	18.37	20.23	Increase (2.95)
	Max. strain (%)	4.52	4.40	4.68	Increase (3.54)

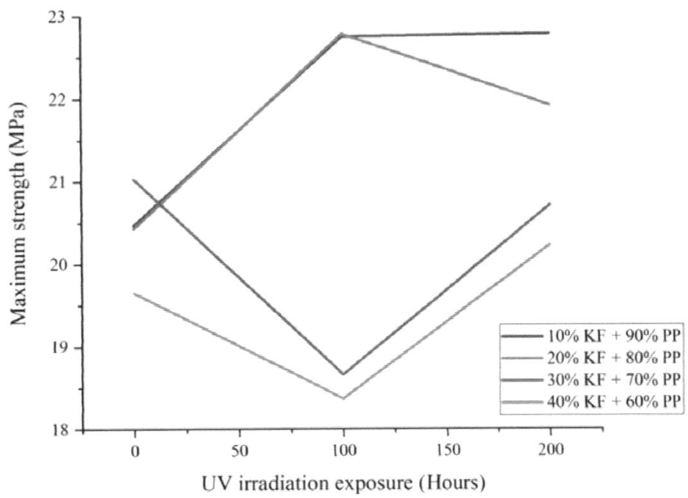

Figure 1. Maximum strength (MPa) of 10% KF + 90% PP, 20% KF + 80% PP, 30% KF + 70% PP and 40% KF + 60% PP samples of UV irradiation exposure (hours).

Table 3 shows the result average of combination kenaf fiber and peat soil with polypropylene upon UV irradiation exposure. Figure 2 shows the increment value of maximum strength starting from 0 to

200 hours of UV exposure. This increments due to the addition of combustion peat soil at 600°C. Contrast unexposed with UV irradiation, the value of tensile strength for 30% KF +70% PP and 40% KF + 70% PP are degraded with the addition of peat soil. The more significant change that occurs in tensile strength along the exposure of 200 hours of UV is the combination of 40% KF + 60% PP + 6% PS with the increment of 6.31%. After 200 hour of UV exposure, the highest maximum strength is 30% KF + 70% PS + 3% PS with the value 21.22 MPa of maximum strength because it has the least composition of kenaf fiber and the addition combustion of peat soil. Peat soil itself similar to any lignocellulose content when damp, but desiccation reduces its hydrophilicity which is drying ultimately leads to hydrophobic peat and high resistance to rewetting.

Table 3. Result average of strength of combination kenaf fiber and peat soil with polypropylene upon UV irradiation exposure.

Samples	Mechanical Properties	Hour			Decrease/Increase (%)
		0	100	200	
30% KF + 70% PP + 0% PS	Max. strength (MPa)	21.03	18.66	20.71	Decrease (1.52)
	Max. strain (%)	5.21	5.13	5.18	Decrease (0.58)
30% KF + 70% PP + 3% PS	Max. strength (MPa)	20.11	20.77	21.22	Increase (5.52)
	Max. strain (%)	5.45	5.37	4.82	Decrease (11.56
30% KF + 70% PP + 6% PS	Max. strength (MPa)	19.98	20.57	20.66	Increase (3.4)
	Max. strain (%)	5.31	4.97	4.46	Decrease (16.01)
40% KF + 60% PP + 0% PS	Max. strength (MPa)	19.65	18.37	20.23	Increase (2.95)
	Max. strain (%)	4.52	4.40	4.68	Increase (3.54)
40% KF + 60% PP + 3% PS	Max. strength (MPa)	18.98	19.31	19.32	Increase (1.79)
	Max. strain (%)	4.25	4.15	3.78	Decrease (11.07)
40% KF + 60% PP + 6% PS	Max. strength (MPa)	18.69	20.43	19.87	Increase (6.31)
	Max. strain (%)	4.58	4.42	3.86	Decrease (15.82)

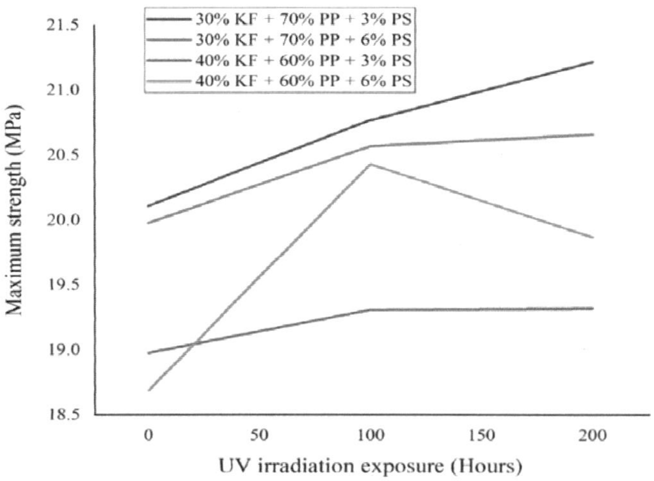

Figure 2. Maximum strength (MPa) combination of 30% KF + 70% PP, 40% KF + 60% PP with different percentage of PS added exposed with UV irradiation (hours).

4. Conclusion

In general, the physical characterization of wood polymer composites endurance to ultraviolet exposure has been successfully studied in this research. During this research, the use of kenaf fiber as a wood substitute in WPC has gained immense popularity, as it provides significant advantages such as being recyclable, environmentally friendly, and often low cost. Similar to the previous research, peat soil's effect reinforced with kenaf fiber and polypropylene was seen to contribute to new composite growth.

From the result obtained, the physical and mechanical properties of kenaf fiber reinforce with polypropylene changed with the exposure to 200 hours of UV. For the first batch that only contains kenaf fiber with 10%, 20%, 30%, 40% and 90%, 80%, 70%, 60% polypropylene, the result highest value of tensile strength after 200 hours UV exposure is 22.78 MPa that obtain from the least composition of kenaf fiber with polypropylene. This because the less composition kenaf fiber in a polymer composite, the lowest degradation occurs by increasing the UV exposure time. The result also shows the increment of composition peat soil into PP and Kenaf Fiber before UV exposure also does not increase the mechanical properties of WPC, but for the outside uses, the combination of peat soil is suitable material since it can withstand with UV exposure. This can be seen after 200 hours UV exposure will increase a little bit maximum tensile strength of WPC which is the highest tensile strength is 21.22 MPa that obtain from the 30% KF + 70% PP + 3% PS.

Acknowledgments

The authors acknowledge the supports from the Ministry of Education (MOE) through the Fundamental Research Grant Scheme (FRGS/1/2018/TK03/UTHM/03/17) and Universiti Tun Hussein Onn Malaysia for sponsoring this work under research grant No. K067.

5. References

[1] Mohammed L, Ansari M N M, Pua G, Jawaid M and Saiful Islam M 2015 *Int. J. Polym. Sci.* **2015** 243947
[2] Depuydt D E C, Soete J, Asfaw Y D, Wevers M, Ivens J, Van Vuure A V 2019 *Compos. -A: Appl. Sci. Manuf.* **119** 48
[3] Gardner D J, Han Y and Wang L 2015 *Curr. Forestry Rep.* **1** 139
[4] Mohd Joharudin N F, Abdul Latif N, Mustapa M S, Badarulzaman N A and Mahmod M F 2020 *J. Adv. Res. Fluid Mech. Therm. Sci.* **68(1)** 125
[5] Rady M H, Mustapa M S, Wagiman A, Al-Alimi S, Shamsudin S, Lajis M A, Mansor M N and Harimon M A 2020 *Int. J. Integr. Eng.* **12(3)** 82
[6] Mohd Masirin M I, Ali A S B, Mustapa M S, Abdul Rahman R, Wagiman A and Aziz M I 2020 *Mater. Sci. Forum* **975** 197
[7] Mohd Joharudin N F, Abdul Latif N, Mustapa M S, Mansor M N, Siswanto W A, Murugesan J and Yusof F 2019 *Materialwiss. Werkstofftech.* **50(3)** 283
[8] Chen Y, Stark N M, Tshabalala M A, Gao J and Fan Y 2016 *Materials (Basel)* **9(8)** 610
[9] BallaV K, Kate K H, Satyavolu J, Singh P and Tadimeti J G D 2019 *Compos. B: Eng.* **174** 106956
[10] Husin M M, Mustapa M S, Wahab M S, Tajul Arifin A M, Mohd Masirin M I and Jais F H 2017 *AIP Conf. Proc.* **1846(1)** 020008
[11] Husin M M, Mustapa M S, Wahab M S, Hashim M Y and Jais F H 2017 The morphology and flexural stress of surface treated and untreated kenaf fiber reinforced polypropylene composites. *Int. J. Mech. Mechatronics Eng.* **17(1)** 56120168
[12] Husin M M, Mustapa M S, Wahab M S, Tajul Arifin A M, Ganasan R and Jais F H 2017 *Mater. Sci. Forum* **909** 94
[13] Hashim U R, Jumahat A, Jawaid M, Dungani R and Alamery S 2020 *Polymers* **12** 2621
[14] Mohammed M, Rozyanty A R, Beta B O, Adam T, Osman A F, Salem I A S, Dahham O S, Al-Samarrai M N and Mohammed A M 2017 *J. Phys. Conf. Ser.* **908** 012003

Study on Wear Debris Characterization of Polycarbonate Urethane (PCU) as a Bearing of Artificial Hip Joint

W D Lestari[1*], R Ismail[2], J Jamari[2], and A P Bayuseno[2]

[1]Department of Mechanical Engineering, Faculty of Engineering, University of Pembangunan Nasional "Veteran" Jawa Timur, 60294, Indonesia
[2]Department of Mechanical Engineering, Faculty of Engineering, Diponegoro University, 50275, Indonesia

*wahyu.dwi.tm@upnjatim.ac.id

Abstract. As with all artificial joints, wear debris is of particular concern due to its effect on both implant life and the in vivo biological reactions that can occur. The purpose of the research is to study debris characterization of PCU. Wear particle is produced from testing the PCU material using a pin on disc wear tester within 50000 cycles. This study showed that the PCU wear debris gotten from the simulator had various different shapes, including laminar and spherical types. The morphology of worn surface and wear debris analysis showed that wear mechanism of PCU were fatigue wear. Thus we conclude that PCU is expected to be a lifetime implantation of artificial joint.

1. Introduction

Artificial joint replacement is the main method of therapy that can effectively relieve joint pain and restore joint function for severe osteoarthritis of the hip and knee joints [1]. During the use of hip and knee implants there is relative contact and movement between the metal surface and the polyethylene which results in the wear fragments of the polyethylene. These wear fragments migrate to the tissue around the implant, which can lead to a biological response to inflammation and bone osteolysis. Furthermore, osteolysis will lead to implant loosening and failure, so revision surgery is required. The failure of these implants made researchers work hard to search for new implant materials with better tribological characteristics. This search aims to reduce the presence of wear and tear during the implant wearing process.

A new type of polymer material has been developed recently which is given the name PCU. The PCU causal performance has been observed through testing with a pin on disc (POD) tester. The presence of wear flakes with different morphologies will have a major impact on the biocompatibility of the implant material. The wear debris characterization resulted after the sliding process. From the results of this characterization will be used as an experimental basis for PCU material which will be applied in the future medical world. These wear specimens can be collected as a result of wear testing via a simulator or from revision hip or knee implant surgery.

There are many studies that focus on the quantification, characterization and compatibility of wear fragments resulting from hip implants from both in vitro and in vivo wear [2][3][4]. One study found the amount of UHMWPE wear particles in the 0.1–1 μm range.

Cherif et al. [5] analyzed the alumina-zirconia material produced during an abrasion-induced wear test. These wear debris contributes to the redistribution of the load applied to more points of contact.

Based on the observations it is shown that debris reduces surface porosity at low loads and in turn reduces friction. At high loads, due to large plastic deformations and cracks, the surface becomes rough which causes increased friction. Colombie et al. [6]have studied the role of wear debris in fretting and concluded that the temperature rise during fretting wear is negligible and that the wear on the first body is governed by the shaping and maintenance of the third body (debris) and the abrasiveness of the third body.

Investigate the characteristics of the wear debris including quantity, size and morphology, then the wear status of mechanical components can be predicted [7–10]. There are various methods to characterize wear flakes, one of which is by using a SEM machine. This machine will provide excellent qualitative information about wear particles, and with the help of computer vision techniques can also provide quantitative analysis. The aim of this research is to characterize the wear debris of PCU materials for medical devices such as bearings in artificial hip joints.

2. Materials and Method

Particles are obtained from the results of wear testing using a pin on disc wear tester. The applied load was 500 gram for 50000 cycles. The test was carried out in dry conditions. The PCU disc were 100 mm in diameter and 10 mm in thickness. The counterparts to the PCU disc samples were pin made from SS 316 L. The counterparts had a 6 mm diameter and a smooth hemisphere on one side. They were 60 mm in length. To investigate the microstructure features on the fracture interface of PCU, the shape and size of the wear debris was characterized with scanning electron microscope (SEM). EDX analysis was used to determine the chemical composition of the particles. SEM and EDX tests were carried out using the JSM-6510LA tool which is a high performance multipurpose SEM with high precision optics.

3. Results and Discussion

The PCU material that has gone through the wear testing process is then carried out by microstructural observations of the wear parts using SEM. The results of the SEM test results for PCU disc specimen wear in the condition without lubrication can be seen in Figure. 1. Based on the image, a wear analysis was performed on the specimen. From the SEM test results, it can be seen that several phenomena occur in the wear track area of the specimen. There is a phenomenon of plastic deformation and some debris on the surface of the wear track. In addition, on the wear track surface there are a lot of abrasive particles because they have crossed the plastic deformation limit. The phenomenon of the appearance of abrasive particles indicates the occurrence of the abrasive wear process.

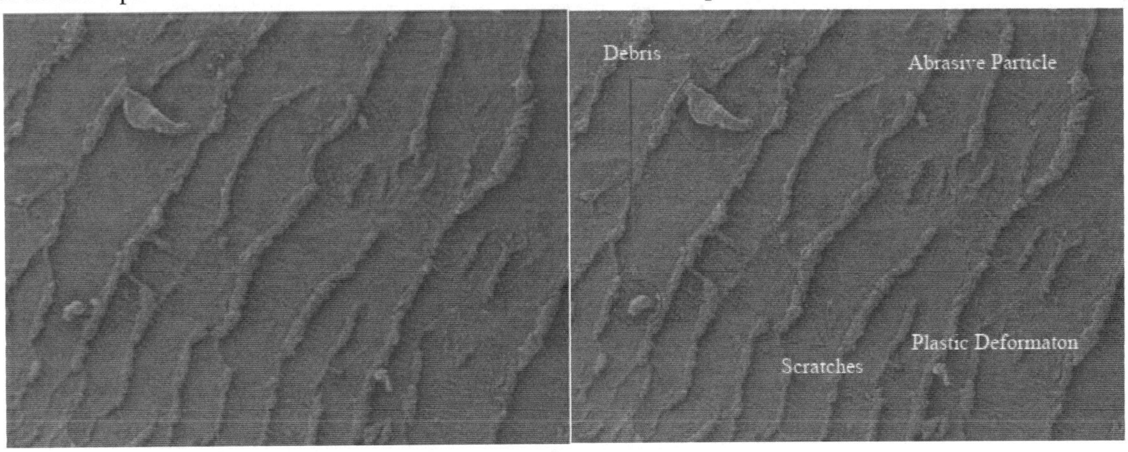

Figure 1. SEM results of wear track disc specimens without lubrication

During the pin on disc testing process, there is a phenomenon of reducing the material in the form of wear particles. The wear particle results from the pin on disc test for the PCU disc specimen were also carried out by SEM testing to determine the shape and size of the wear particle and EDX testing to find out what content was in the PCU disc specimen after pin on disc testing. Figure.2 until Figure.4 showed that the PCU wear debris obtained from the pin on disc wear tester

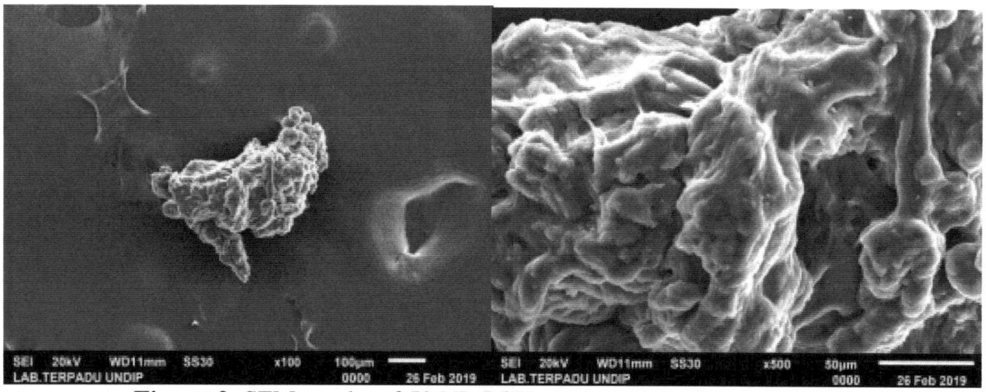

Figure 2. SEM results of 50 NL PCU wear particle disc specimens

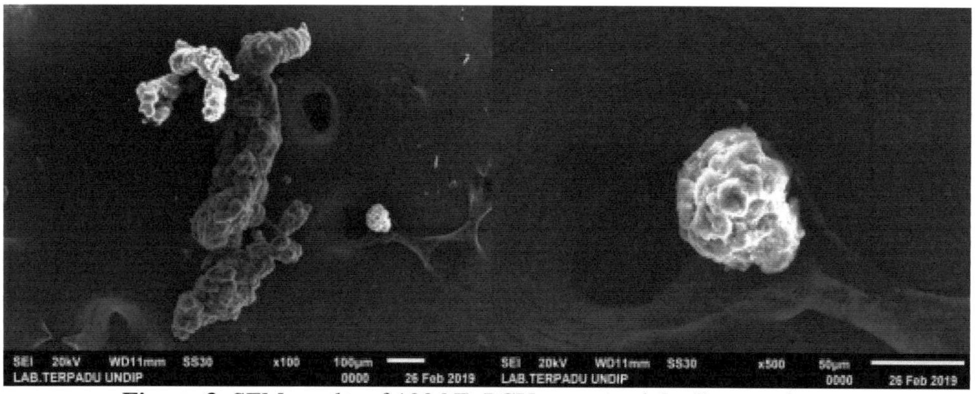

Figure 3. SEM results of 100 NL PCU wear particle disc specimens

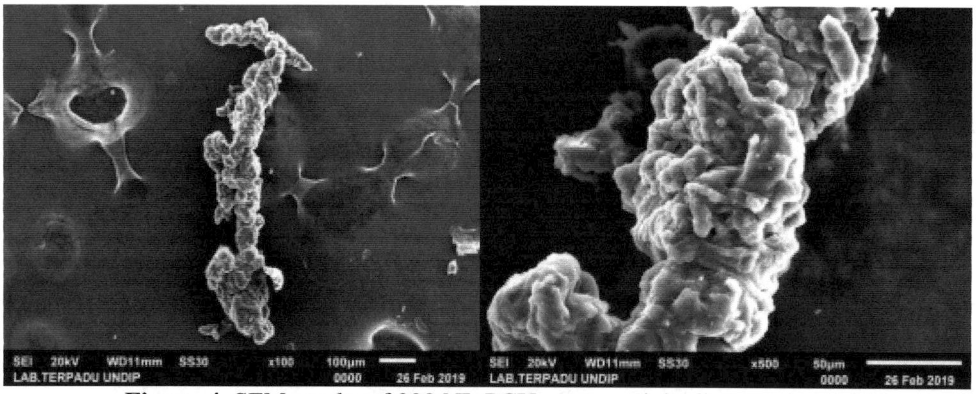

Figure 4. SEM results of 200 NL PCU wear particle disc specimens

Based on the results of the SEM test for the wear particles, then processing is carried out using measurement software to obtain size data from the wear particle. The results of SEM wear particle testing using measurement software are presented in Table 1. From the measurement results, it was found that the size of the wear particle of each PCU disc has a length and width of hundreds of micrometers. The forms of wear particles that are formed are laminar and spherical types. This type of wear particle indicates the type of fatigue wear. This happens because of the continuous loading received by the PCU material when it undergoes a wear test of 50,000 cycles.

EDX testing is also carried out to determine the content contained in the wear particle of the PCU disc specimen. Overall, the EDX test results can be seen in the Table 2. In the EDX test on all specimens, there are no Fe, Cr and Ni elements which are the constituent elements of the AISI 316L stainless steel pin material attached to wear particle PCU specimen disc. Therefore it can be said that the contact mechanism in this pin on disc test is not at risk of producing highly reactive free radicals that can damage DNA, proteins and other cell components [11].

Table 1. Size of wear particle from SEM results

No.	Spesimens	Length (μm)	Wide (μm)
1.	50 NL	447.545	466.015
2.	100 NL	76.823	70.121
3.	200 NL	820.521	299.346

Table 2. EDX wear particle test results for each specimen

No.	Compounds	50 NL	100 NL	200 NL
1.	C	C (100%)	C (99.14%)	C (99.27%)
2.	O	-	O (0.17%)	O (0.20%)
3.	Cu	-	Cu (0.68%)	Cu (0.43%)
4.	Al	-	-	Al (0.10%)
5.	CuO	-	CuO (0.86%)	CuO (0.54%)
6.	Al_2O_3	-	-	Al_2O_3 (0.19%)

4. Conclusion

Based on the results of SEM, it shows that the wear particle for PCU disc specimens has a size ranging from 50 to hundreds of micrometers, so the wear particle for this type of PCU disc specimen is the type of fatigue. The EDX results show that there are no harmful substances or compounds in the PCU wear particle disc specimens. Further research is needed regarding pin on disc wear testing of polycarbonate urethane (PCU) materials in dry lubrication conditions in various variables, types of pin material and cycles. Thus, we consider the PCU have the potency to take place of UHMWPE, certainly, further research need to be done in the future.

5. References

[1] Bullock MW, Brown ML, Bracey DN, Langfitt MK, Shields JS, Lang JE 2017 *J Arthroplasty* vol 32(4) pp 1067–73.
[2] Garcin S, Fouvry S, Heredia S 2015 *Wear* vol 330–331 pp 145–59.
[3] Yue T, Wahab MA 2016 *Materials* vol 9(7).
[4] Jin X, Sun W, Shipway PH 2016 *Tribol Int* vol 102 pp 392–406.
[5] Cherif K, Gueroult B, Rigaud M 1997 *Wear* vol 208(1–2) pp 161–8.
[6] Vincent L, Berthier Y, Floquet A, Godet M 1984 *J Tribol.* vol 106(2) pp 192–200.
[7] Yuan W, Chin KS, Hua M, Dong G, Wang C 2016 *Mech Syst Signal Process* vol 72–73 pp 346–58.
[8] Wang J, Bi J, Wang L, Wang X 2018 *Mech Syst Signal Process* vol 100 pp 863–76.
[9] Zhu X, Du L, Zhe J 2017 *Mech Syst Signal Process* vol 83 pp 296–304.
[10] Hong W, Wang S, Liu H, Tomovic MM, Chao Z 2017 *Mech Syst Signal Process* vol 82 pp 1–12.
[11] Chen Q, Thouas GA 2015 *Mater Sci Eng R* vol 87 pp 1–57.

Journal of Physics: Conference Series **2065** (2021) 012005 doi:10.1088/1742-6596/2065/1/012005

Efficiency improvement of solar cells by $CaAlSiN_3:Eu^{2+}$ and $Y_2O_3:Eu^{3+}$ phosphors

Shaoqiang Huang, Chaogang Lou*, Han Diao, Zhaoyong Wang and Yunzhen Yin

School of Electronic Science and Engineering, Southeast University, Nanjing, China

lcg@seu.edu.cn

Abstract: $CaAlSiN_3:Eu^{2+}$ (CASN) and $Y_2O_3:Eu^{3+}$ (YO) phosphors are mixed into ethyl vinyl-acetate (EVA) to form luminescence down-shifting (LDS) films which are used to improve crystalline silicon solar cells by converting the photons with the wavelength from 200 nm to 500 nm into red light. Experimental results show that the conversion efficiency of the solar cells is enhanced from 19.61% to 20.00%. The improvement is better than that with only CASN phosphors in EVA films. The reason is the high quantum yield of YO phosphors in the wavelength range 250 nm-350 nm, which offsets the poorer performance of CASN phosphors in this wavelength range.

1. Introduction

Although the conversion efficiency of solar cells has been continuously increasing in recent years [1-2], they are still suffering the decrease in the performance after packaging because of the increased surface reflection and the absorption in the encapsulation materials [3-5]. Luminescent down shifting (LDS) materials doped into encapsulation materials is considered to be a possible way to improve this problem [6]. Although the quantum yield of LDS materials is less than 1, they can still enhance the efficiency of solar cells by converting short wavelength photons into longer wavelength photons to which the solar cells have better response [7].

Lots of LDS materials have been investigated for solar cells and can be separated into three main categories: quantum dots, organic dyes and rare-earth phosphors [8-10]. Compared with quantum dots and organic dyes, rare-earth phosphors are the most applicable LDS materials for the solar cells due to their large Stokes shift (>200 nm), high quantum yield and low cost [11].

Many researchers have applied LDS phosphors on the solar cells. Although the performance of the cells are improved, these phosphors have limited effect because their narrow exaction band [12-15]. The $CaAlSiN_3:Eu^{2+}$ (CASN) phosphors have wider exaction band and emit red light where the crystalline silicon (c-si) solar cells exhibit a better response, so it is a good candidate for solar cells [16-17]. However, the quandum yield of CASN phosphors become lower (< 50%) below 300 nm, which restricts the improvement of the solar cells.

In this work, we choose the $Y_2O_3:Eu^{3+}$ (YO) phosphors which have high quantum yield ($\approx$ 100%) below 300 nm to offset the poor performance of CASN phosphors in the wavelength range below 300 nm [18]. We introduce the mixture of CASN and YO phosphors into ethylene vinyl acetate (EVA) to form LDS films which are used to encapsulate the solar cells. This process is similar to that of production lines and convenient to be applied in photovoltaic industry. To make a comparison, we also make EVA films with only CASN phosphors and EVA films without any phosphor.

2.Experiment

The CASN and YO phosphors are purchased from Grirem Advanced Materials Company, and the c-si solar cells (the size is 2 cm X 2 cm) are provided by Maodi Solar Technology Company. The LDS films with CASN and YO phosphors were prepared as follows. The EVA, CASN phosphors, YO phosphors were mixed according to the mass ratio of 1: 0.015: 0.015 ratios. After that, the mixture was heated to 180°C, and then pressed into the films with the thickness of 0.3 mm in a vacuum chamber. The LDS films with only CASN phosphors was made through the similar procedure, where the mass ratio is EVA: CASN=1:0.015. Then, the LDS films, single c-si solar cells and glasses were packaged together in a vacuum chamber of 0.1 Pa and at temperature 160°C. The schematic structure of the modules is shown in Figure1.

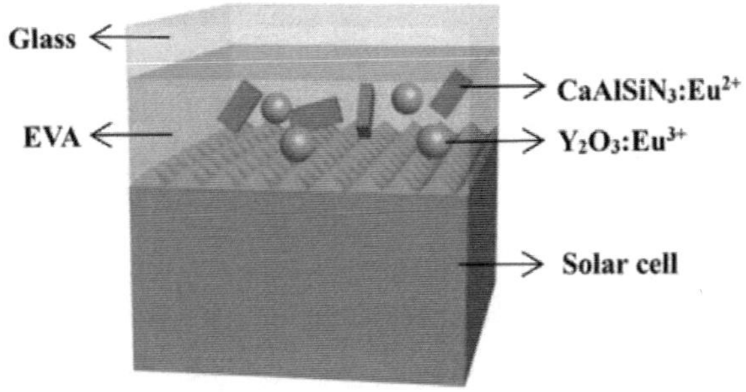

Figure 1. Schematic structure of the solar cell modules with CASN and YO phosphors

The photoluminescence (PL) and photoluminescence excitation (PLE) spectra of CASN and YO phosphors were measured by Ocean Optics MAYA 2000PRO. The absorption of films were tested by SHIMADZU UV-1780. Qtest Station 2000AD produced by Crowntech is used to measure the external quantum efficiency (EQE) and the surface reflectivity of the solar cells. Two light sources, tungsten lamp and deuterium lamp, are used in combination to scan the wavelength range 200-1100 nm. The conversion efficiency was tested by IEC AAA XJCM-9 (GSolar).

To demonstrate the effects of the CASN and YO phosphors on the solar cells, three c-si solar cells have been intentionally selected. The three cells have similar short circuit currents (J_{sc}), open circuit voltages (V_{oc}), fill factors (FF) and conversion efficiency (η). Three packaged solar cells are: the sample S1 does not include any phosphor, the sample S2 includes only CASN phosphors and the sample S3 includes CASN and YO phosphors.

3.Results and discussion

Figure 2 shows the normalized photoluminescence excitation (PLE) and photoluminescence (PL) spectra of CASN, YO phosphors. The CASN phosphors have a broad excitation band from UV region to 590 nm and emits the light with the peak around 605 nm, which is a typical 5d-4f transitions of Eu^{2+} [19]. While, because of $^5D_0 \rightarrow {}^7F_2$ transitions within Eu^{3+}, the YO phosphors exhibits luminescent properties with nearly 100% QE [20]. These results indicates that the mixture of CASN and YO phosphors has an excitation band which is wide enough to cover the low response band of the solar cells.

The performance of the bare solar cells and the packaged cells are summarized in Table 1. In order to reduce the measuring errors, the parameters of each sample have been measured for six times and the average values are given in Table 1. It can be noted that V_{oc} and FF are kept nearly unchanged before and after packaging. The conversion efficiency of S1 (without phosphor) is 19.38%, lower than 19.58% of its bare cell (the relative growth is -1.02%). The conversion efficiency of S2 (with CASN phosphors) increases from 19.62% of the bare cell to 19.86% of the packaged cell (relative growth is 1.22%), and the relative growth of S3 (with CASN and YO phosphors) is 1.98%. Clearly, the introduction of CASN and YO phosphors gives a best improvement of the solar cells.

To explain these experimental results, the EQE and the reflection of the cells are measured, as shown in Figure 3. For the packaged cells, the EQE is slightly increased and the reflection is decreased in the region of 300-400 nm after packaging. This is because packaged cell has the structure of air-glass-EVA-cells which forms an index gradient and reduces the reflection in the short wavelength range[21-22]. In the middle and long wavelength ranges, the reflectivity of packaged cells increases and their EQE is lowered slightly. These results agree well with previous reports [23-24].

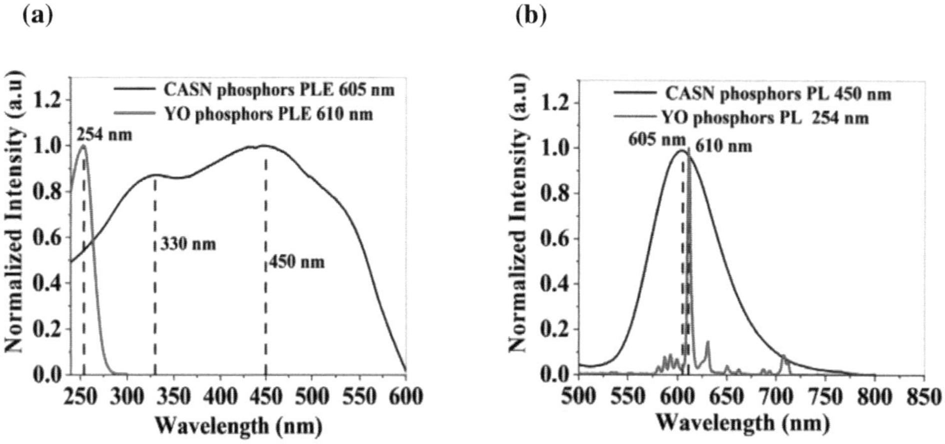

Figure 2. (a) Normalized photoluminescence excitation (PLE) spectra of CASN and YO phosphors monitored at 605 nm and 610 nm, respectively. (b) Normalized photoluminescence (PL) spectra for CASN and YO phosphors excited by 450 nm and 254 nm, respectively.

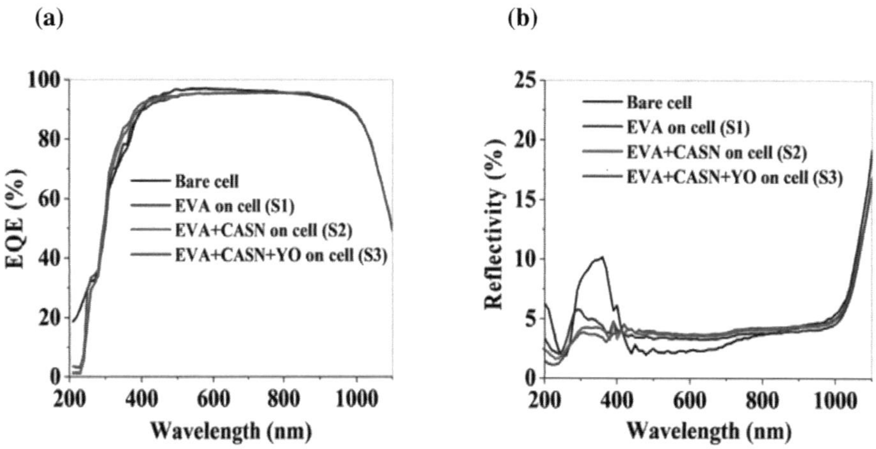

Figure 3. (a) External quantum efficiency. (b) Reflectivity of solar cells

By comparing packaged cell of S1 (without phosphor) and S2 (with CASN phosphors), it can be seen that the introduction of CASN phosphors brings a rise in the conversion efficiency by 2.24% relatively. This can be explained by the LDS effect of the CASN phosphors. The CASN phosphors absorb the short lights in the range of 250-500 nm and produce red emission peaking at 605 nm and lower the reflection.

As for the packaged cell S3 (with CASN and YO phosphors), its increase in the conversion efficiency is lager than that of S2 (with CASN phosphors). The introduction of YO phosphors improve the packaged cells through two possible mechanisms: one is that YO phosphors increase the absorption of the light in the range of 250-300 nm; other is that YO phosphors scatter the light of 300-500 nm, which increases the absorption of the light by CASN phosphors.

Table 1. Performance parameters of bare cells and packaged cells.

Sample		J_{SC} (mA/cm^2)	V_{OC}(V)	FF	η (%)	ΔJ_{sc} (%)	Δη (%)
S1	Bare Cell	38.62	0.655	0.774	19.58	-1.17	-1.02
	EVA on Cell	38.17	0.656	0.774	19.38		
S2	Bare Cell	38.62	0.653	0.778	19.62	1.19	1.22
	EVA + CASN on Cell	39.08	0.654	0.777	19.86		
S3	Bare Cell	38.64	0.654	0.776	19.61	2.25	1.98
	EVA + CASN +YO on Cell	39.51	0.654	0.774	20.00		

4. Conclusions

This paper proposes a method for improving the efficiency of crystalline solar cells by using $CaAlSiN_3:Eu^{2+}$ (CASN) and $Y_2O_3:Eu^{3+}$ (YO) phosphors. The YO phosphors improve the luminescence down-shifting effect of CASN phosphors in the wavelength range below 300 nm and increasing the absorption. Although introduction of the YO and CASN phosphors brings higher reflection, the improved spectral response results in a slight increase of the energy conversion efficiency from 19.61% to 20.00%.

5.References

[1] Green M A, et al. Solar cell efficiency tables (version 57). *Progress in Photovoltaics: Research and Applications*, 2020, **29**: 3-15.

[2] Yoshikawa K, et al. Silicon heterojunction solar cell with interdigitated back contacts for a photoconversion efficiency over 26%. *Nature Energy*, 2017, **2**: 1552-1576.

[3] Machado M, et al. Optical model for multilayer glazing systems: Experimental validation through the analytical prediction of encapsulation-induced variation of PV packaged cells efficiency. *Solar Energy*, 2016, **135**: 77-83.

[4] McIntosh K R, et al. Increase in External Quantum Efficiency of Encapsulated Silicon Solar Cells from a Luminescent Down-Shifting layer. *Progress in Photovoltaics: Research and Applications*, 2009, **17**: 191-197.

[5] Klampaftis E, et al. Improvement in multi-crystalline silicon solar cell efficiency via addition of luminescent material to EVA encapsulation layer. *Progress in Photovoltaics: Research and Applications,* 2011, **19**: 345-351.

[6] Klampaftis E, et al. Enhancing the performance of solar cells via luminescent down-shifting of the incident spectrum: A review. *Solar Energy Materials and Solar Cells*, 2009, **93**: 1182-1194.

[7] Huang X, et al. Enhancing solar cell efficiency: the search for luminescent materials as spectral converters. *Chemical Society reviews*, 2013, **42**: 173-201.

[8] Cao Y, et al. Down-Shifting and Anti-Reflection Effect of $CsPbBr_3$ Quantum Dots/Multicrystalline Silicon Hybrid Structures for Enhanced Photovoltaic Properties. *Nanomaterials*, 2020, **10**: 775.

[9] Pardo Perdomo A, et al. Luminescent down-shifting film based on optimized mixture of organic dyes for improving the performance of $P_3HT:PC_{61}BM$ photovoltaic devices. *Journal of Applied Physics*, 2020, **128**: 35502.

[10] Cardoso M A, et al. Solar spectral conversion based on plastic films of lanthanide-doped ionosilicas for photovoltaics: Down-shifting layers and luminescent solar concentrators. *Journal of Rare Earths*, 2020, **38**: 531-538.

[11] Correia S F H, et al. Luminescent solar concentrators: challenges for lanthanide-based organic–inorganic hybrid materials. *Journal of Materials Chemistry A*, 2014, **2**: 5580-5596.
[12] Levchuk I, et al. Industrially scalable and cost-effective Mn^{2+}doped $Zn_xCd_{1-x}S/ZnS$ nanocrystals with 70% photoluminescence quantum yield, as efficient down-shifting materials in photovoltaics. *Energy & Environmental Science*, 2016, **9**: 1083-1094.
[13] Chander N, et al. Improving the short wavelength spectral response of silicon solar cells by spray deposition of YVO_4: Eu^{3+} downshifting phosphor nanoparticles. *IEEE J. Photovolt.* 2015, **5**: 1373-1379.
[14] Shao G, et al. Enhancing the efficiency of solar cells by down shifting YAG: Ce^{3+} phosphors. *Journal of Luminescence*, 2015, **157**: 344-348.
[15] Karunakaran S K, et al. Efficiency improvement of Si solar cells by down-shifting Ce^{3+} -doped and down-conversion Ce^{3+} -Yb^{3+} co-doped YAG phosphors. *Solar Energy*, 2019, **188**: 45-50.
[16] Uheda K, et al. Luminescence properties of a red phosphor, $CaAlSiN_3$: Eu^{2+}, for white light-emitting diodes. *Electrochemical and Solid State Letters*, 2006, **9**: H22-H25.
[17] Kim H S, et al. Luminescence properties of $CaAlSiN_3$:Eu^{2+} phosphor prepared by direct-nitriding method using fine metal hydride powders. *Journal of Alloys and Compounds*, 2015, **633**: 97-103.
[18] Ronda C R. Recent achievements in research on phosphors for lamps and displays. *Journal of Luminescence*, 1997, **72**: 49-54.
[19] Ueda J, et al. Thermal quenching mechanism of $CaAlSiN_3$: Eu^{2+} red phosphor. *Bulletin of the Chemical Society of Japan*, 2018, **91**: 173-177.
[20] Meltzer R S, et al. Dependence of fluorescence lifetimes of Y_2O_3: Eu^{3+} nanoparticles on the surrounding medium. *Physical Review B*, 1999, **60**: R14012.
[21] Cao H, et al. A Packaging Method to Improve Monocrystalline Silicon Solar Cells with YAG: Ce Phosphors, *2019 3rd International Conference on Circuits, System and Simulation (ICCSS). IEEE*, 2019: 225-228.
[22] Shao G, et al. Luminescent down shifting effect of Ce-doped yttrium aluminum garnet thin films on solar cells. *Applied Physics Letters*, 2015, **107**: 253904.
[23] Liu J, et al. Improving spectral response of monocrystalline silicon photovoltaic modules using high efficient luminescent down-shifting Eu^{3+} complexes. *Progress in Photovoltaics: Research and Applications*, 2013, **21**: 668-675.
[24] Tahhan A, et al. Photovoltaic cells energy performance enhancement with down-converting photoluminescence phosphors. *International Journal of Energy Research*, 2015, **39**: 1616-1622.

The Application of Brain-computer Interface (BCI) based Functional Electrical Stimulation (FES)

Tianhang Liu

RDFZ Chaoyang Branch school, Beijing, China, 100028

liutianhang@rdfzcygj.cn

Abstract. Rehabilitation medicine has developed rapidly in recent years. Brain computer interface and functional electrical stimulation are very cutting-edge technologies in this field. Because brain computer interface provides a real-time operation platform for patients to operate their limbs according to their intention for functional electrical stimulation, the research on BCI based FES has gradually increased in recent years. This paper discusses the current development status and technical application of FES and BCI. The research status of BCI based FES is discussed, and the existing problems of various research are summarized. According to the research findings, this field is a new technology with great application prospects in the field of modern rehabilitation engineering.

1. Introduction

Functional electrical stimulation (FES)'s principle is to stimulate nerves or muscles through low-frequency pulse current, cause muscle contraction, and then improve or restore the function of stimulated muscles or muscle groups. In the past decades, FES technology has been applied in many different aspects of the medical field. Such as the recovery of bladder function after spinal cord injury, the treatment of respiratory and swallowing disorders caused by stroke, the improvement of motor function in patients with paralysis, and the well-known cardiac pacemaker. However, it should be pointed out that FES treatment is most suitable for the human body with a complete nerve conduction pathway, that is, muscle dysfunction caused by central nervous system (brain and spinal cord) injury, while it has a poor effect on muscle weakness caused by peripheral nerve injury without complete nerve conduction pathway [1], but it can be used to promote the regeneration of injured peripheral nerve and prevent denervated atrophy of skeletal muscle. BCI (Brain computer interface) is a new communication and control technology established between the human brain and external devices (such as computers or other electronic devices). It does not depend on the normal output pathway of brain-peripheral nerve-muscle. The expression of will and the manipulation of equipment can be realized by collecting and analyzing EEG signals.

The control of its stimulation signal restricts the further development of FES. Because if we can't find the appropriate stimulation signal, FES can't achieve a good treatment effect, and its motion control of the residual limb can only be carried out according to the preset mode, and can't carry out the limb motion control at will in real-time according to the wishes of the patient. Thus, BCI technology provides a very good operation interface, the combination of BCI and FES technology in the treatment of patients with nerve or muscle injury is the latest application in the field of rehabilitation. I summarize all the application fields and technical problems of BCI based FES so far.

It provides a reference for future researchers in this field.

2. BCI based FES technique

2.1. Working system of BCI based FES

Generally, the working system of BCI based FES is divided into two modules. The first part is BCI, the module includes collecting signals from the skin surface and amplifying them for signal conditioning. Then, it classifies the required signals through filters, and finally records and transmits these signals. The second module is FES. Signals produced by the BCI will be transmitted to FES equipment. These FES devices will read the signal and generate corresponding electrical pulses. These electrical pulses are sent to the muscle surface, and the muscle will contract.

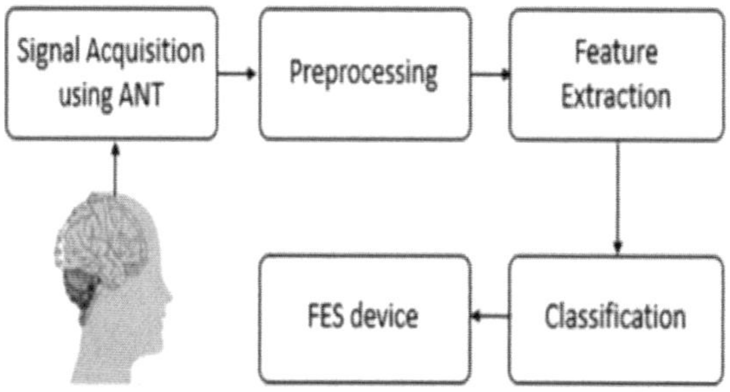

Figure 1. Block system for the working system of BCI based FES [2]

In 2003, the combination of brain computer interface and functional electrical stimulation was applied for the first time, which brought hope to paralyzed patients and attracted the attention of many researchers. The basic principle of the system is: the brain computer interface is used as the control source to analyze people's intentions; Functional electrical stimulation technology is used as the driving source to stimulate muscles and make paralyzed limbs move [3].

EEG is the most important technique used in brain computer interface to collect brain signals caused by neuron activity. Its non-invasive interface, economy, convenience, and high time resolution make it popular for all researchers. Lin etal pointed out that Brain computer interface based on EEG can be divided into four categories: steady-state visual evoked potential (SSVEP), slow cortical potential (SCP), P300 and Mu and beta rhythm of motor imagery [4]. Therefore, all papers related to BCI based FES so far use SSVEP and p300 as their method to collect EEG signals.

2.1.1. SSVEP based FES. Based on the urgent practical application, Gollee et al consider the brain computer interface based on SSVEP, which has higher information transmission speed, simpler training and relatively small individual differences compared with other EEG acquisition methods [5]. SSVEP is a low-level response with a constant frequency [6]. If there are a large number of visual stimuli that oscillate at different frequencies, the selected stimulus can be detected by matching the SSVEP response frequency to one of the stimulus frequencies. In the literature [7, 8, 9], the SSVEP response is usually described as reliable.

In Lin's paper, patients should look at the black-and-white flashing boxes with different frequencies on the screen according to their wishes, to select different upper extremity movement modes. The function of brain computer interface of steady-state visual evoked potential is to collect EEG signals of the scalp when the eyes are looking at a flashing area, extract and classify the signals, identify the flashing frequency of the patient's staring area, and then translate the patient's intention [4].

Figure 2 is the black and white flashing boxes [4]. 12.5hz flickering corresponds to hand grasp action, 7.5hz flickering corresponds to elbow extension, 8.33hz strobe corresponds to hand open action, 6.82hz strobe corresponds to elbow flex action, and 9.37hz corresponds to idle state. The narrow bar next to each white flashing area is an identification energy bar. On the one hand, it is used to identify whether the visual attention area is consistent with the result. On the other hand, the energy bar changing from green to red is also a switch to trigger the interface between BCI and FES, that is, after reaching a certain threshold, the corresponding area setting an intention is transmitted to the electric stimulator through the interface, so as to realize the switching between different electric stimulation modes, control different actions.

Figure 2. Black and white flashing boxes [4]

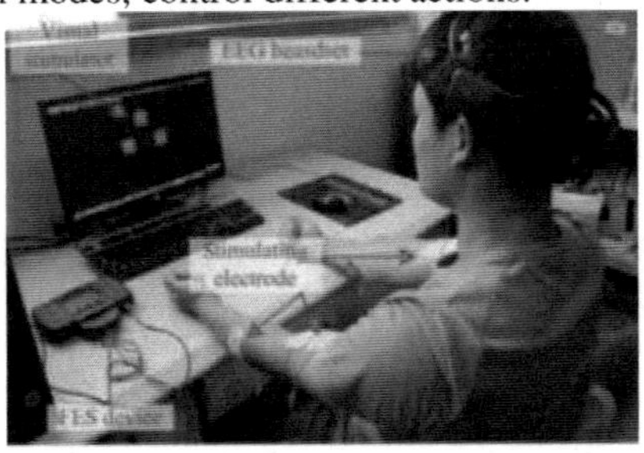

Figure 3. Experimental setup to achieve FES [10]

Another system, shown in Figure 3, uses SSVEP based BCI to achieve functional electrical stimulation. Zhao et al. designed a similar system with the former one [10]. In order to achieve the rehabilitation of the upper extremity, the integrated system has three upper limb operation modes: fast mode, medium speed mode and low speed mode. The three flickering squares are mapped to three motion modes. The square of 20 Hz represents the fast mode, the square of 15 Hz represents the medium speed mode, and the square of 12 Hz represents the slow mode. In addition, a square without flickering in the center corresponds to an idle state.

2.1.2. P300 based FES. Visual or auditory mental images and external stimuli can generate EEG signals. By using auditory stimuli (such as headphones) to generate P300, the user's brain supplement activities are more focused, and it is not like visual stimuli that distract patients from hill [11]. This signal control method also does not require the extensive training required by mental imaging techniques, and the resulting equipment is not user-specific.

A method combining PCA and ICA technology has been proven effective in processing data, but it has not yet been implemented in the auditory P300 BCI. Alexander et al. conducted experiments on this issue. They conducted 6 experiments on 15 participants. Six of the 180 trials were performed on 15 subjects (right-handed men between the ages of 21 and 30) using the high-resolution system [12].

In the end, it was found that the accuracy of EEG signal with p300 as the carrier was as high as 78%, but the training time was long and it was not easy to experiment.

2.2. Classification of BCI based FES

2.2.1. Upper extremity. The technology of BCI based FES to help the rehabilitation of upper limb movement is very promising, such as upper limb hemiplegia caused by stroke. The combination of FES and BCI can help elbow or shoulder voluntary flexion and extension, forearm rotation or finger movement. The following are the relevant literature, as well as their research methods and processes.
Nourhan et al. conducted an experiment on recovering the activities of paralyzed upper limbs through BCI based FES [2]. The system uses a certain method. ANT neuro headphones collect data, carry out signal processing and feature extraction, and connect FES stimulators. However, in this experiment, the FES module selected transcutaneous electrical stimulation (TENS). But at the same time, this method uses the medium frequency modulation composite waveform, which can stimulate the deeper muscle tissue and make the muscle contract. The experiment successfully achieved arm extension and bending.

Muller-Putz et al. conducted a case study of connecting an EEG-based BCI with an implantable FES device freehand system. In the study, the patients were first trained for a short period of time in order to make them produce obvious EEG patterns by imagining the movements of the disabled hand. This mode is further classified by BCI, and then the output signal generated by the BCI simulated shoulder joystick is provided to the freehand system of different gripping stages. With continuous imaging, the patient can move a simple object from one place to another [13]. Zhou Peng et al. used the left and right hand motion imaginary potential as the input signal of BCI, and used it in conjunction with FES to develop an intelligent rehabilitation system for paralyzed patients to control the movement of their residual limbs according to their own motion wishes. This system can bypass the damaged nerve pathway in the body of the paralyzed patient, directly convey the person's movement intention to the FES instrument through the external pathway, stimulate the corresponding neuromuscular, and complete the patient's direct control of the residual limb. Preliminary experiments have shown that this system can analyze people's movement intentions with a correct rate of more than 95%, control the FES instrument to complete the predetermined stimulation tasks, and resume hand grasping actions [14]. Tan et al. completed the rehabilitation training experiment of wrist and finger of stroke patients with BCI based NMES. The motor image of μ rhythm (8-12Hz) was used in this experiment. When a person wants to exercise, mu rhythm will decay, which is used to trigger NMES on the wrist [15].

However, the experiment requires a lot of training for participants to maintain a high degree of concentration. Goniometer at the joint to provide feedback to determine the completion of NMES. Once NMES is activated, BCI will immediately stop detecting brain waves. Until the goniometer detects that the arm is fully extended, the electrical stimulation stops.

2.2.2. Lower extremity. Lower extremity has more restrictions when designing the whole BCI based FES system compared with upper extremity. Because EEG signals can be extremely unstable because the large movement degree when walking or standing. In order to get stable EEG signals, active shielding electrodes needed [16]. Due to the high difficulty, there is less research on the system serving the lower limbs.

An et al realized ankle rotation by stimulating foot dorsiflexion with BCI based FES, which provided a rehabilitation way for those people with chronic gait impairment caused by foot ptosis [16]. But different from other experiments, volunteers need to accept the computer model prediction first. Five healthy volunteers participated in the experiment. The reaction delay was about 1.4-3.1 seconds, but all volunteers successfully achieved dorsiflexion. When one lower limb performs 10 alternating ankle neutral and dorsiflexion positions, it can trigger the lower limb with BCI-FES on the other side to step on dorsiflexion, which indicates that BCI-FES may be an effective rehabilitation method to solve foot ptosis caused by stroke and brain trauma. It is proved in this paper that BCI and FES is a novel and feasible treatment for patients with lower limb paralysis. Chung et al. conducted EEG-BCI-FES ankle dorsiflexion training for 5 days in 5 stroke patients, the results showed that the score of Berg Balance Scale and walking speed were significantly improved [17]. In addition, he also compared the patients with foot ptosis with FES training alone and BCI FES training respectively. The results showed that the curative effect of the latter was significantly better than the former, indicating that the application of BCI can effectively improve the rehabilitation of patients' lower limb function.

Christine et al. conducted an experiment to control functional electrical stimulation to walk on the ground based on a brain-computer interface [18]. The experiment recruited a patient suffering from spinal cord injury. The patient was first trained in BCI. In order to evaluate the performance of BCI, they used a laser rangefinder gyroscope and video recording data. Followed by FES training until the patient can complete the ground walk completely autonomously. Finally, it is the BCIFES system experiment. The step rate and current stimulation amplitude are programmed into the microcontroller in advance, so that the left and right steps can be automatically cycled. Participants also need to participate in the suspended walking experiment and the ground walking experiment. In the end, the study showed that participants can initially operate under a suspension, then transfer their skills to walking on the ground and achieve a high level of control. The research provides a recovery treatment for paraplegic patients.

3. Existing technical problems

Due to the short development time of BCI based FES, the research and technology in this area are not very complete. Previous studies have pointed out the existing problems in the experimental process and the aspects that can be improved. But the most important thing is that the system is based on BCI and FES, so the development of BCI and FES has a great influence on the perfection of this system.

Peter et al. pointed out that the movement of artificial limb and the increase of muscle and eye activities lead to the application of neural artificial limb, which may cause artifacts in EEG, which will have a bad impact on feature extraction and FES signal transmission [19]. At the same time, many experiments are recruiting healthy volunteers. The differences of brain waves between those with SCI and healthy volunteers may lead to accuracy of the experiment.

Taehoon et al. pointed out that their current research only applied a simple action sequence to participants, which should be completed by themselves [20]. Then, although they determined that the effect of BCI FES on the arm of patients with paraplegia was obvious within 4 weeks, there was no long-term monitoring or investigation effect.

Christine et al. pointed out in the experiment that the error rate of volunteers during walking on the ground is higher than the discontinuous movement of pausing walking [18]. This performance is due to the increase of EEG noise caused by the movement of volunteers, such as the movement of center of gravity and unbalanced shaking of posture. The second point is that the current noninvasive BCI FES system is very cumbersome. In addition, the system has a great impact on the comfort of other parts of volunteers, affecting the activities of brain and nerves and the movement of parts.

For SSVEP based BCI, Zhao et al. showed that SSVEP can only deal with decentralized intentions [10]. The conversion between continuous control and different motion modes cannot be realized. Due to too many unstable factors of EEG signals, such as human factors such as concentration difference and individual difference of subjects, it will directly affect the recognition rate of EEG signals, and further affect the response speed of the whole BCI based FES system, so some subjects may feel uncomfortable such as dazzling, which will also increase the false recognition rate.

Tan et al. emphasized that the use of EEG will bring some disadvantages [15]. The impedance matching phase takes a long time, and it will always lead to brain fatigue and affect volunteers' attention. Although there were no other side effects, participants usually felt headache after the experiment.

4. Conclusion

This paper discusses the current technical status of BCI and FES, as well as the research status of BCI based FES. Although the application of BCI based FES in limb control and medical rehabilitation is only in the preliminary stage, most of the literature shows that the system is very feasible. It is found that there are fewer studies on BCI based FES for lower limbs. First, there is a stronger demand for recovering the movement of paralyzed upper limbs, while the movement of patients with lower limb paralysis can be completed with the help of auxiliary equipment such as wheelchairs; second, the movement of lower limbs (such as walking) will interfere with electroencephalogram (EEG), so special brain computer interface equipment is required, such as active shielded electrode [16]; third, the position of the cerebral cortex controlling the walking movement of the lower limbs is deep, and the collected EEG signal information is relatively weak, which brings difficulties to the later signal processing and recognition.

In view of the feedback from previous studies, future studies should be more inclined to optimize the experimental scheme and reduce the impact of external factors on patients' brain activities; try to use patients with SCI to carry out the experiment; BCI based FES at present, the experimental equipment is too heavy, and the device should be simplified; Improve the response rate of BCI and FES.

Acknowledgements

I would like to thank Prof Andreas Demosthenous from UCL to provide pre-knowledge about implantable and wearable medical device and several literature related to this topic. And thanks to all teachers, family and friends for their support and guidance on this paper.

5. References

[1] Bogataj, Uros, Gros, et al. The rehabilitation of gait in patients with hemiplegia: A comparison between conventional therapy [J] Physical Therapy, 1995.

[2] Abduallatif N A, Elsherbini S G, Boshra B S, et al. Brain-Computer Interface controlled functional electrical stimulation system for paralyzed arm [C] 2016 8th Cairo International Biomedical Engineering Conference (CIBEC). IEEE, 2016, pp. 3.

[3] Pfurtscheller G. Thought--control of functional electrical stimulation to restore hand grasp in a patient with tetraplegia [J] Neuroscience Letters, 2003, 351(1): 33-36.
[4] Yao L, Zhang D, Huang G, et al. Using SSVEP based brain-computer interface to control functional electrical stimulation training system [C] IEEE. 2011, pp.324-326.
[5] Gollee H, Volosyak I, Mclachlan AJ, et al. An SSVEP-Based Brain－Computer Interface for the Control of Functional Electrical Stimulation [J] IEEE Transactions on Biomedical Engineering, 2010, 57(8): 1847-1855.
[6] Galloway N R. Human Brain Electrophysiology: Evoked Potentials and Evoked Magnetic Fields in Science and Medicine [J] British Journal of Ophthalmology, 1990, 74(4).
[7] Wang Y, Gao X, Bo H, et al. Brain-Computer Interfaces Based on Visual Evoked Potentials [J] IEEE Engineering in Medicine & Biology Magazine the Quarterly Magazine of the Engineering in Medicine & Biology Society, 2008, 27(5): 64.
[8] Friman O, Luth T, Volosyak I, et al. Spelling with Steady-State Visual Evoked Potentials [C] International IEEE/EMBS Conference on Neural Engineering. IEEE, 2007.
[9] Gao X, Xu D, Cheng M, et al. A BCI-based environmental controller for the motion-disabled [J] IEEE Transactions on Neural Systems and Rehabilitation Engineering, 2003, 11(2): 137-140.
[10] Chu Y, Zhao X, Han J, et al. SSVEP based brain-computer interface controlled functional electrical stimulation system for upper extremity rehabilitation [C] IEEE International Conference on Robotics & Biomimetics. IEEE, 2014.
[11] Hill N J, Lal T N, Bierig K, et al. An Auditory Paradigm for Brain--Computer Interfaces [C] Advances in Neural Information Processing Systems 17 [Neural Information Processing Systems, NIPS 2004, December 13-18, 2004, Vancouver, British Columbia, Canada], 2004.
[12] ASJ Bentley, CM Andrew, LR John. An offline auditory P300 brain-computer interface using principal and independent component analysis techniques for functional electrical stimulation application [C] International Conference of the IEEE Engineering in Medicine & Biology Society. IEEE, 2008.
[13] Graupe, Daniel. An overview of the state of the art of noninvasive FES for independent ambulation by thoracic level paraplegics [J] Neurological Research, 2002, 24(5): 431-442.
[14] Zhou P, Cao H B, Xiong Y, et al. Design of intelligent rehabilitation system based on Brain-Computer Interface [J] Computer Engineering & Applications, 2007.
[15] Tan, Kong, Shee, et al. Post-acute stroke patients use brain-computer interface to activate electrical stimulation [J] IEEE, 2010.
[16] An H D, Wang P T, King C E, et al. Brain-Computer Interface Controlled Functional Electrical Stimulation System for Ankle Movement [J] Journal of NeuroEngineering and Rehabilitation, 2011, 8(1):49.
[17] Chung E J, Kim J H, Park D S, et al. Effects of brain-computer interface-based functional electrical stimulation on brain activation in stroke patients: a pilot randomized controlled trial [J] Journal of Physical Therapy Science, 2015, 27(3): 559-562.
[18] The feasibility of a brain-computer interface functional electrical stimulation system for the restoration of overground walking after paraplegia [J] Journal of NeuroEngineering and Rehabilitation, 2015, 12(1):1-11.
[19] Boord P, Ba Rriskill A, Craig A, et al. Brain-computer interface - FES integration: Towards a hands-free neuroprosthesis command system [J] Neuromodulation Journal of the International Neuromodulation Society, 2010, 7(4): 267-276.
[20] Kim T H, Kim S S, Lee B H. Effects of Action Observational Training Plus Brain-Computer Interface-Based Functional Electrical Stimulation on Paretic Arm Motor Recovery in Patient with Stroke: A Randomized Controlled Trial [J] Occupational Therapy International, 2016, 23(1): 39.

Journal of Physics: Conference Series **2065** (2021) 012007 doi:10.1088/1742-6596/2065/1/012007

A Demodulation Circuit For Analog Front End Of Passive Tag Chip

Qinglong Li, Yong Xu*, Qiao Li, Kun Peng, Xian Zhang

School of Communication Engineering, Army Engineering University of PLA, Nanjing, China

xu_yong99@163.com

Abstract. The demodulation circuit designed in this paper is suitable for the analog front end of passive UHF RFID tag chip, which can handle ASK signals with large changes in amplitude, modulation depth and signal frequency. Its performance meets the requirements of standards ISO/IEC 18000-6C and GB/T 29768-2013. Envelope detection circuit and limiter circuit are simple in structure and do not consume power. The comparison reference voltage is taken according to the average value of the envelope high and low levels, and is less affected by the dynamic changes of the input signal. Changing the width-to-length ratio of the MOSFETs in the feedback path of the comparator can adjust the hysteresis, with strong noise suppression and controllable sensitivity. The demodulator is implemented with TSMC 0.18 μm standard CMOS process. The simulation results show that the ASK signal modulation depth that the demodulator can handle is as low as 30%, and the maximum pulse width demodulation error is only 0.43%.

1. Introduction

Radio frequency identification (RFID) technology relies on radio waves to realize non-contact data exchange and target identification between reader and tag chip. The ultra-high frequency (UHF, 300 MHz ~ 3 GHz) radio frequency system has the characteristics of long identification distance, fast transmission speed and small antenna size[1]. The current international mainstream standards and national standards for UHF RFID systems are ISO/IEC 18000-6C and GB/T 29768-2013, respectively. It can be known from the two standards that the system forward link signal is a wave modulated by amplitude keying (ASK) with a modulation depth of 30% to 100%, and the baseband signal frequency is 40 kHz to 160 kHz.

The structure of the tag chip demodulator is generally a low-pass filter followed by a comparator. Its comparison reference voltage takes half of the envelope peak voltage. This circuit has a simple structure but cannot demodulate ASK signals with a modulation depth of less than 50%. Some schemes are voltage-compensated envelope detector followed by peak detection circuit. Although this type of circuit improves the demodulation sensitivity, the structure of the detection circuit is complex and requires a working power supply[2]. This paper designs a UHF RFID demodulator with no power consumption for the first two sub-circuits and based on average detection. It can handle RF signals with large changes in amplitude, modulation depth and signal frequency. The circuit performance meets the requirements of ISO/IEC 18000-6C and GB/T 29768-2013.

2. Design of UHF RFID Signal Demodulator

The structure of the UHF RFID signal demodulator designed in this paper is shown in Figure 1.

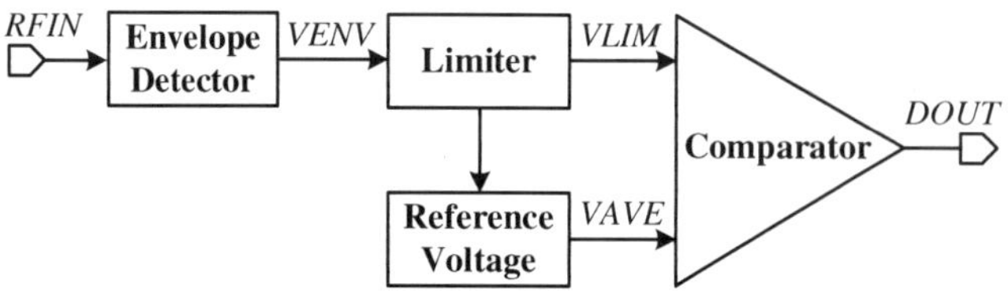

Figure 1. Demodulator of UHF RFID signal

The received radio frequency signal $RFIN$ modulated by amplitude keying (ASK) is transmitted to the envelope detector to extract the envelope $VENV$ of the modulated signal. The limiter circuit can avoid circuit abnormalities caused by large envelope signals. The limiting voltage $VLIM$ is divided into two lines: one is connected to the comparator; the other is transmitted to the reference voltage generator to generate the comparison reference value $VAVE$. The comparator compares $VLIM$ and $VAVE$ to determine the high or low level, and finally outputs the digital signal $DOUT$.

2.1 Envelope Detector

As shown in Figure 2, a better scheme to extract ASK signal envelope is voltage multiplier plus low-pass filter[3]. The circuit structure is very simple, does not require additional power supply, does not consume power, and is suitable for use in passive tag chips.

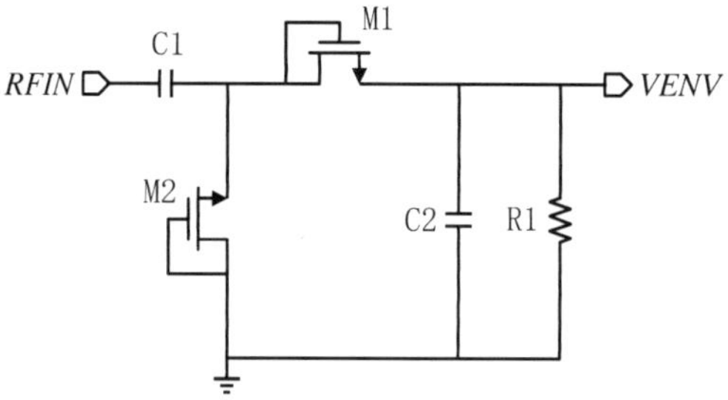

Figure 2. Envelope detector circuit

Capacitors C1, C2 and MOSFETs M1 and M2 in Figure 2 form a one-stage voltage doubler rectifier circuit. When the value of $RFIN$ is less than the threshold voltage V_{TH} of the MOSFET, that is, $-V_{TH} < RFIN < V_{TH}$, both M1 and M2 are cut off. When $RFIN$ is in the negative half cycle and $RFIN < -V_{TH}$, M2 is turned on, and the current flows from Ground through M2 to charge C1. Assuming that the peak value of $RFIN$ is V_M, the voltage on C1 will be $V_{C1} = -(V_M - V_{TH})$ at this time. As the input voltage rises, M2 will be cut off, and according to the conservation of charge, the voltage on C1 will remain unchanged. When $RFIN > V_{TH}$, M1 turns on and $RFIN$ starts charging C2. When $RFIN$ reaches the peak value of the positive half cycle, the ideal voltage value of C2 should be:

$$V_{C2} = V_M - [-(V_M - V_{TH})] - V_{TH} = 2(V_M - V_{TH}) \quad (1)$$

When $RFIN$ gradually drops from the peak value of the positive half cycle to $RFIN < V_{TH}$, M1 is cut off and C2 discharges through R1. Since the charging time constant $\tau_1 = R_{on1} \cdot C2$ is much smaller than the discharge time constant $\tau_2 = R1 \cdot C2$, V_{C2} is better maintained near the peak value of the positive half cycle of $RFIN$. V_{C2} is $VENV$.

The simulation result of the envelope detection circuit given in the second row of Figure 6 shows: the passive and power-free voltage doubler rectifier and low-pass filter structure successfully extract the envelope of the radio frequency signal.

2.2 Limiter

When the distance to the reader is closer, the energy received by the tag is larger. The limiter circuit can avoid the abnormality of the subsequent circuit caused by the large envelope signal. As shown in Figure 3, the entire limiter circuit is not connected to a working power source and has no energy consumption.

Diode-connected N-MOSFETs M1~M4 are connected in series to form a voltage detection channel. Resistor R1, R2 and N-MOSFETs M5, M6 form a charge leakage path[4].

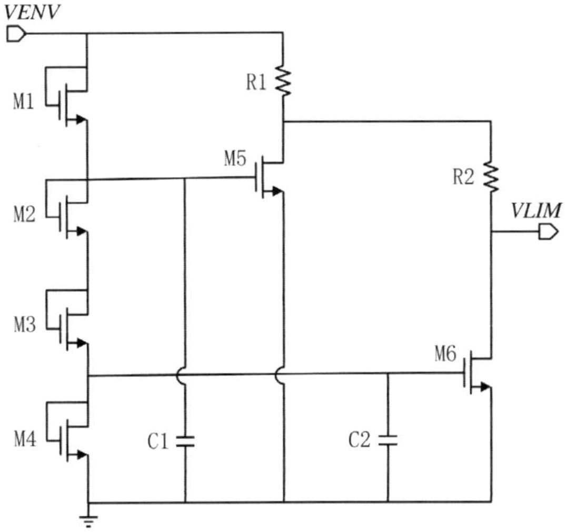

Figure 3. Limiter circuit

When $VENV$ is low, the gate-source voltage V_{GS5} of M5 is less than its threshold voltage V_{TH5}, so V_{GS6} must be less than V_{TH6}. At this time, M5 and M6 are both in the cut-off state and the output voltage $VLIM$ is equal to the input voltage $VENV$.When $V_{GS5} > V_{TH5}$ and $V_{GS6} < V_{TH6}$, M5 is on and M6 is off. The resistance R1 and the equivalent resistance r_{O5} of M5 divide the input voltage. At this time, the output voltage is:

$$VLIM = VENV \cdot \frac{r_{O5}}{R_1 + r_{O5}} \tag{2}$$

When $VENV$ rises continuously so that $V_{GS6} > V_{TH6}$, both M5 and M6 are turned on. The output voltage is:

$$VLIM = VENV \cdot \frac{r_{O5} \| (R_2 + r_{O6})}{R_1 + r_{O5} \| (R_2 + r_{O6})} \cdot \frac{r_{O6}}{R_2 + r_{O6}} \tag{3}$$

Figure 4 shows the simulation results of the limiter circuit. When the input voltage is 0~1.5 V, the output voltage is approximately equal to the input voltage. When the input voltage is 1.5~8 V, the output voltage is limited to 1.8 V.

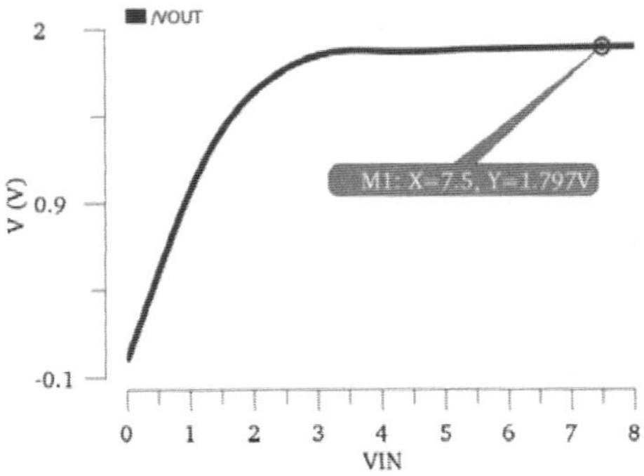

Figure 4. Simulation results of limiter

2.3 Reference Voltage Generator

Divide $VLIM$ into two lines: one is connected to the comparator; the other is connected to the reference voltage generator to generate the average voltage $VAVE$. If you want to demodulate radio frequency signals with different amplitudes and different modulation depths, the reference voltage used for comparison should be dynamically determined according to the high and low levels of $VLIM$. The reference voltage generator based on peak value V_{peak} and valley value V_{vall} detection satisfies this requirement well. Its circuit structure is shown as in Figure 5[5]. The operational amplifier in the circuit can adopt a general operational amplifier structure.

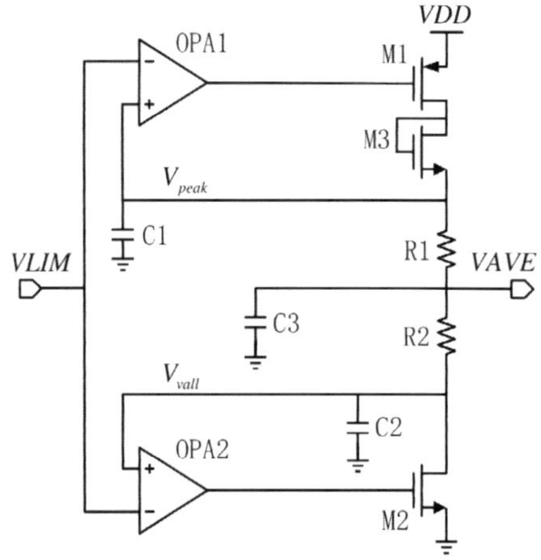

Figure 5. Reference voltage generator circuit

The operational amplifiers OPA1 and OPA2 form a voltage follower through a deep negative feedback loop. OPA1, P-MOSFET M1 and capacitor C1 jointly detect the peak value V_{peak}. N-MOSFET M3 adjusts the amount of current in the channel. OPA2, N-MOSFET M2 and capacitor C2 can detect the valley value V_{vall} of $VLIM$. Equivalent resistors R1 and R2 divide the peak and valley voltages to determine the average voltage $VAVE$, that is:

$$VAVE = \frac{1}{2}(V_{peak} + V_{vall}) \tag{4}$$

At the initial moment, the voltages on capacitors C1, C2 and C3 are all zero. When the input terminal $VLIM$ is high level, the voltage of the inverting terminal of OPA1 and OPA2 is greater than their in-phase terminal. Therefore, both operational amplifiers output low level. M1 turns on and M2 turns off. VDD charges C1, C2 and C3 through M1. Because OPA1 is a voltage follower, the C1 voltage V_{peak} will be equal to the input voltage $VLIM$. M2 cut-off makes the negative feedback loop of OPA2 open and cannot form a voltage follower. The C2 voltage V_{vall} is divided by R1 and R2. Because the resistances of the equivalent resistors R1 and R2 are relatively large, the charging time constant is large, so V_{vall} rises slowly. According to the KVL theorem, the C3 voltage $VAVE$ conforms to equation (4).

The input voltage $VLIM$ changes from high level to low level. The voltage V_{peak} of the energy storage capacitor C1 is greater than $VLIM$, so that the operational amplifier OPA1 outputs a high level. So M1 is cut off. The negative feedback loop of OPA1 is open, making V_{peak} no longer follow $VLIM$ changes. C2 voltage V_{vall} is also greater than $VLIM$, making OPA2 output a high level. Then M2 is turned on. C1, C2 and C3 are discharged through M2. Since the discharge time constants of C1 and C3 are much greater than the signal period, the input level $VLIM$ has already become high before V_{peak} and $VAVE$ drop in time. Because the on-resistance of M2 is small, the discharge time constant of C2 is small. So V_{vall} quickly drops to a low level. Due to the voltage follower principle, V_{vall} remains near the low level. Similarly, the equivalent values of R1 and R2 show that $VAVE$ satisfies equation (4).

The simulation result of the reference voltage generator is shown in the third row of Figure 6: The actual $VAVE$ (dashed line) generated is successfully placed between the high and low levels of $VLIM$.

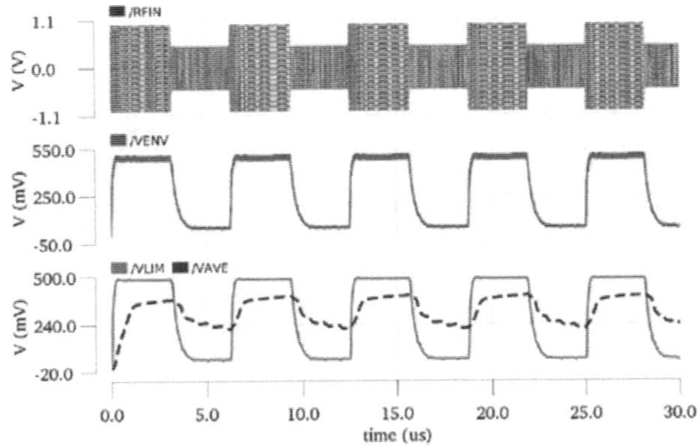

Figure 6. Simulation result of reference voltage generator

2.4 Hysteresis Comparator

Ripple voltage input to an ordinary single-threshold comparator will produce back and forth jitter at the output. The hysteresis comparator has the ability to suppress noise whose amplitude is within the range of hysteresis ΔV. The structure of the hysteresis comparator designed in Figure 7 is a two-stage amplifier followed by a two-stage inverter[6].

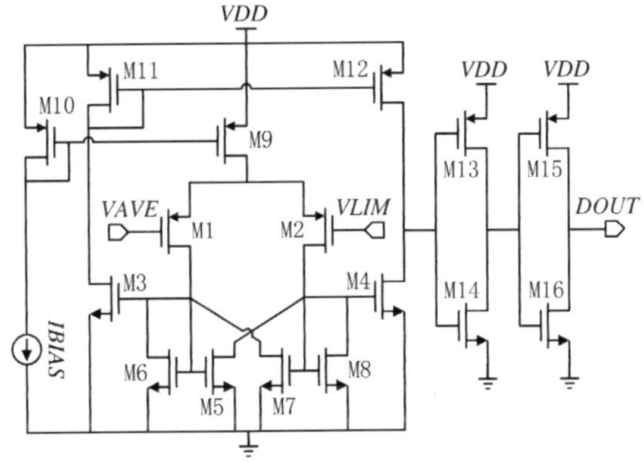

Figure 7. Hysteresis comparator circuit

The differential input MOSFETs M1 and M2 have the same specifications. Now assume that $VAVE$ is grounded. When the input voltage $VLIM$ is far less than 0, the relationship between the gate-source voltage of M1 and M2 is: $|V_{GS1}| \ll |V_{GS2}|$. Then M2 turns on and M1 turns off. The turned-on M2 causes M7 and M8 to turn on, but M1 cutoff makes M5 and M6 cut off and makes M7 in the deep linear region. So there is $I_9 = I_2 = I_8$ (I_2, I_8, and I_9 are the drain-source currents of MOSFETS M2, M8, and M9, respectively.). At this time, M3 is off and M4 is on, and the comparator outputs a low level.

The input voltage $VLIM$ gradually increases from low to high, causing $|V_{GS2}|$ to continue to decrease. Then M1 gradually leaves the cut-off state. The ultimate role of M7-M8 mirroring is to:

$$I_7 = \frac{(W/L)_7}{(W/L)_8} \cdot I_8 \qquad (5)$$

$(W/L)_7$ and $(W/L)_8$ in formula (5) are the conductive channel width-to-length ratios of MOSFET M7 and M8, respectively. When the rising $VLIM$ makes M1 gradually turn on and makes I_1 increase to the mirror value I_7, the output of the comparator switches from low to high. Therefore, When $I_1 = I_7$, the input voltage difference $(VLIM - VAVE)$ is the positive turning point. This point is the upper threshold voltage V_+, so:

$$V_+ = VLIM - VAVE = V_{GS2} - V_{GS1} \quad (6)$$

From $I_1 = I_7, I_2 = I_8, I_9 = I_1 + I_2$ and equation (5), the following equation can be derived:

$$I_1 = \frac{I_9}{1+\frac{(W/L)_8}{(W/L)_7}} \quad (7)$$

$$I_2 = \frac{I_9}{1+\frac{(W/L)_7}{(W/L)_8}} \quad (8)$$

From equations (6) to (8) and the current formula of the saturation region of the P-MOSFET (where μ_P is the surface electron mobility of the P-MOSFET, C_{OX} is the gate capacitance per unit area):

$$V_+ = \sqrt{\frac{2I_9}{\mu_P C_{OX}(W/L)_1}} \cdot \left(\frac{1}{\sqrt{1+\frac{(W/L)_8}{(W/L)_7}}} - \frac{1}{\sqrt{1+\frac{(W/L)_7}{(W/L)_8}}}\right) \quad (9)$$

Similarly, the lower threshold voltage can be derived as:

$$V_- = \sqrt{\frac{2I_9}{\mu_P C_{OX}(W/L)_1}} \cdot \left(\frac{1}{\sqrt{1+\frac{(W/L)_5}{(W/L)_6}}} - \frac{1}{\sqrt{1+\frac{(W/L)_6}{(W/L)_5}}}\right) \quad (10)$$

The bias current I_9 of the circuit is provided by the current mirror M9. If the aspect ratio of M1 (M2) has been determined, the upper and lower threshold voltages and hysteresis can be determined by changing the aspect ratios of M5, M6, M7 and M8. Figure 8 shows the simulation result of the hysteresis response of the comparator: the designed hysteresis of this circuit ΔV is 20 mV.

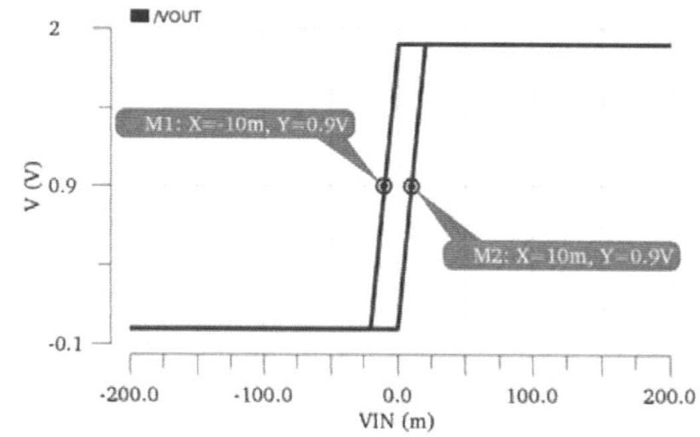

Figure 8. Hysteresis response of comparator

3. Circuit Layout

In order to reduce the mutual interference between circuit devices, each sub-circuit and each amplifier are arranged as independent module units. There are two guard rings around each unit: the inside is the P guard ring connected to the ground, and the outside is the N guard ring connected to the power supply. The combination of inner P and outer N forms a PN junction wall around the outside of the module to protect the inside of the module from external noise. The capacitors, resistors, P-MOSFETs and N-MOSFETs in each module unit should be arranged separately. A P guard ring connected to the ground is arranged around each capacitance area, resistance area and N-MOSFET area. This protects the holes in the P substrate in the ring from being injected by electrons outside the ring. In addition, the P-ring of the N-MOSFET area can also be used as the substrate of the MOSFET. An N-ring connected to the power supply is arranged around the P-MOSFET area: This not only protects the electrons in the N-well in the ring from being absorbed by the holes outside the ring, but also serves as the substrate of the P-MOSFET.

The current mirror controls the current of each branch by changing the width-to-length ratio of each MOSFET. In order to obtain an accurate proportional current, the MOSFET in the current mirror should adopt interdigital structure. Such as hysteresis comparator M5-M6, M7-M8, M9-M10 and M11-M12. Each MOSFET that needs to be matched should have the same channel length. The matching of the two MOSFETs uses the ABBA symmetrical interdigital structure and the three uses ABCCBA.

Because the MOSFET on the differential input terminal is extremely sensitive to signals, the accuracy of the one-dimensional interdigital structure is insufficient. The two-dimensional co-centroid array has better consistency, symmetry, dispersion, compactness and directionality. M1-M2 and M3-M4 in the hysteresis comparator are differential input pairs. The channel lengths of these MOSFETs that need to be matched should be the same. The two-dimensional high-precision matching MOSFET's fingers should be divided into two rows: the layout of the first row is ABBA and the second row is BAAB to form an $\frac{\mathrm{ABBA}}{\mathrm{BAAB}}$ cross-coupled pair.

Figure 9 shows the demodulation circuit layout of the tape-out.

Figure 9. Layout of demodulation circuit

4. System Simulation of Demodulator

The pre-simulation and post-simulation of the whole circuit use cadence spectre. The working power supply of the circuit is 1.8 V during simulation. The carrier of the radio frequency signal is a sine wave of 915 MHz. The simulation conclusion is obtained by observing and measuring the output signal and calculating the maximum deviation. Figure 10 shows the simulation results of ASK signals with three different amplitudes, modulation depths and pulse frequencies. The widths of the three input pulse signals are 3.125us, 6.25us and 12.5us respectively. The actual measured output pulse width is 3.116us, 6.277us and 12.499us respectively. Thus, The pulse width errors are -0.29%, 0.43% and -0.01%, respectively.

Protocols ISO/IEC 18000-6C and GB/T 29768-2013 use PIE and TPP to encode baseband data respectively. Both of these codes can be converted into non-return-to-zero (NRZ). Now arbitrarily give three groups of 8-bit NRZ baseband data, and use different depths to modulate the carrier to generate ASK signals. Use them as incentives for the circuit. The output response is shown in Figure 11: The baseband signal is accurately and completely demodulated by this circuit.

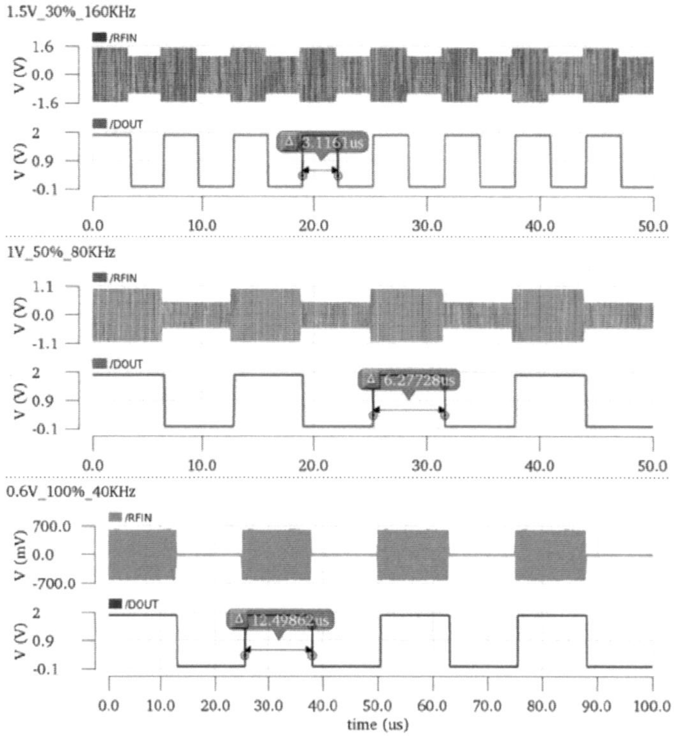

Figure 10. Simulation output of multiple combinations

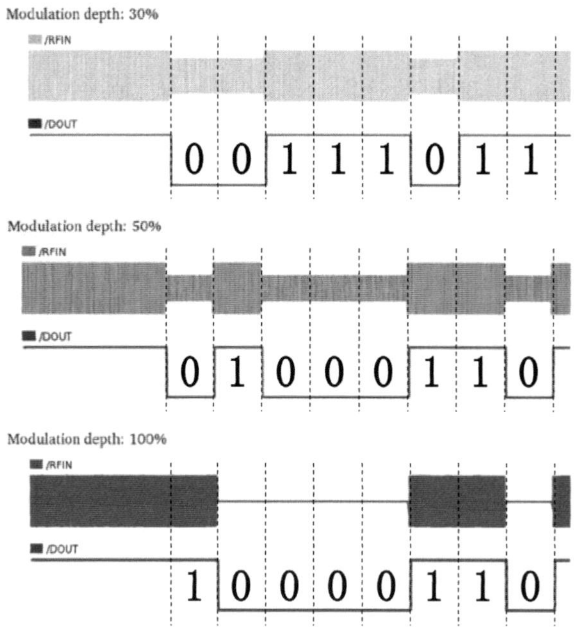

Figure 11. Simulation output of 8-bit random NRZ

The demodulation circuit designed in this paper with good dynamic performance and extremely low demodulation error has been applied to an ultra-low power label chip that has been taped out. The position of the demodulation module in the chip is shown in the photomicrograph in Figure 12.

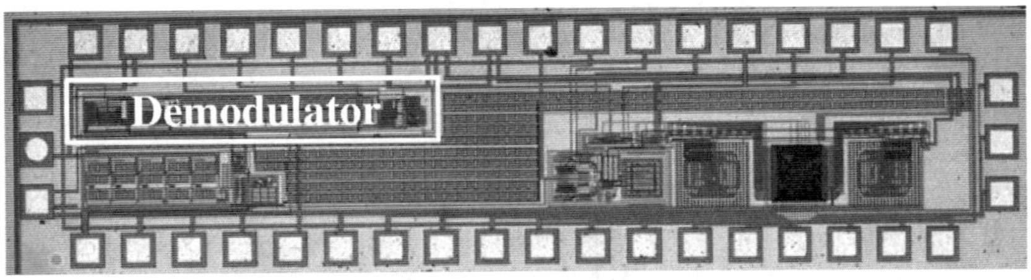

Figure 12. Micrograph of the chip

5. Conclusion

The envelope detection circuit and limiter circuit in this paper do not need power supply and consume no power, so they are suitable for passive tags. The reference voltage generator can generate the corresponding average value reference voltage according to the high and low level of the signal envelope, so its dynamic adaptation performance is better. Finally, after the judgment of a hysteresis comparator with strong noise suppression, it can demodulate ASK radio frequency signals with various amplitudes, various modulation depths and various code rates. Reasonable layout technology minimizes device mismatch and parasitic effects, so the circuit has a good front-end consistency. The circuit is implemented using TSMC 0.18 μm standard CMOS process. The simulation results of ASK signal combinations of various amplitudes, modulation depths and pulse frequencies show that the maximum error of the signal pulse width is only 0.43%. The demodulator can accurately and completely process the PIE and TPP encoding of the protocol ISO/IEC 18000-6C and GB/T 29768-2013. This circuit has been used as a demodulation module in passive tag chips and successfully taped out.

6. References

[1] A. El Boutahiri, M. El Alaoui, K. El Khadiri, A. Tahiri, and H. Qjidaa, "Design of new power generating circuit for passive UHF RFID tag," International Journal of Power Electronics and Drive Systems, vol. 9, no. 3, pp. 1389-1397, 2018.

[2] D. Yongqian, Z. Yiqi, L. Xiaoming, and L. Weifeng, "A UHF RFID chip solution with new oscillator calibration scheme and 16k bits EEPROM," IEICE Electronics Express, vol. 13, no. 13, pp. 20160472-20160472, 2016.

[3] L. Yang, J. Li, M. Tang, L. Cai, J. Li, and M. Zheng, "A high-sensitivity ASK demodulator for passive UHF RFID tags with automatic voltage limitation and average voltage detection," in 2015 IEEE 11th International Conference on ASIC (ASICON), Chengdu, China, 2015, pp. 1-4: IEEE.

[4] S. Haddadian and J. C. Scheytt, "Analysis, Design and Implementation of a Fully Integrated Analog Front-End for Microwave RFIDs at 5.8 GHz to Be Used With Compact MIMO Readers," IEEE Journal of Radio Frequency Identification, vol. 4, no. 4, pp. 476-490, 2020.

[5] Y.-L. Tsai, J.-Y. Chen, B.-C. Wang, T.-Y. Yeh, and T.-H. Lin, "A 400MHz 10Mbps D-BPSK receiver with a reference-less dynamic phase-to-amplitude demodulation technique," in 2014 Symposium on VLSI Circuits Digest of Technical Papers, Honolulu, HI, USA, 2014, pp. 1-2: IEEE.

[6] S. Li et al., "A− 20 dBm Passive UHF RFID Tag IC With MTP NVM in 0.13-μm Standard CMOS Process," IEEE Transactions on Circuits and Systems I: Regular Papers, vol. 67, no. 12, pp. 4566-4579, 2020.

Journal of Physics: Conference Series **2065** (2021) 012008 doi:10.1088/1742-6596/2065/1/012008

Design and Research of Highly Integrated Multi-channel Tile-type T/R Module

Ziran Chen, Zhi Huang*,Qi Zhang and Meng Zhang

General Department, Aerospace Science & Industry Academy of Communication Technology, Chengdu, China

15686796@qq.com

Abstract. This paper introduces a design scheme and key technology of a multi-channel tile-type T/R module. In order to meet the requirements of high reliability, miniaturization, lightweight and multi-functional integration, a vertical interconnection design method of high integrated multi-channel tile-type T / R module is proposed. LTCC technology is used to achieve high-density stacking of radio frequency devices, while bare chips are used to improve system integration and reliability. The module integrates 4 isolated transceiver channels with a size of 40mm×40mm×10mm, which significantly reduces the volume and weight compared to the traditional T/R module structure. The module simulation results prove that the design scheme is easy to implement and fault location maintenance, and the test results meet the design indicators.

1. Introduction

Active phased array antenna has been widely used in military and civilian fields such as radar and communications[1]. The T/R module is the most important component in this type of antenna. The RF signal direction of the traditional brick-type T/R module is parallel to the module installation direction, which has the advantages of simple circuit, mature technology and low cost[2,3]. However, with the increase of the working frequency, the element spacing of the active phased array antenna becomes very small, resulting in a very thin thickness of the brick assembly, which is even difficult to realize in engineering. Moreover, the module is far away from the cold plate, with long heat dissipation path, large thermal resistance and complex heat dissipation structure[4]. At the same time, the installation method determines its large longitudinal size, which makes it difficult to reduce the profile of the entire active antenna. It cannot be used for active phased array antenna applications with low profile requirements such as limited space size, special loading platforms or conformal array[5,6].

Tile-type T/R module was developed earlier in foreign countries, and some progress has been made. Mark S. Hauhe et al. adopted a new high-density package design. The package used multilayer aluminum nitride (AlN) substrates, flipped monolithic microwave integrated circuit (MMIC) chips, coplanar waveguide transmission lines, and solderless fuzz button interconnects[7].

The domestic research on the technical problems and implementation technology of tile-type T/R module started relatively late. Zhiguang Zhang et al. proposed a method to achieve effective vertical interconnection. Microstrip in the cavity is connected to radiating unit through offset strip line, coaxia[8]. Jinzhong Hao et al. proposed a new 3D cubic packaging technology in the highly integrated packaging T/R module. The module has 4 receiver and transmitter channels, and the multifunctional monolithic microwave integrated circuit(MMIC) chip technique and multi- chip module(MCM) technique are also adopted to increase the integration level[9].

The working frequency of the tiled T/R module in this paper is the X-band, which has strict requirements for high reliability, miniaturization, light weight, and multi-function integration. However, the above models have technical problems in heat dissipation structure, engineering application and other aspects, which can't meet the design requirements or engineering batch application.

Based on the above analysis, this paper provides a highly integrated multi-channel tile-type T/R module for the shortcomings of the existing technology. The module integrates multiple RF channels to achieve miniaturization, which can effectively reduce the profile height of the active phased array antenna, and the internal heating device of the T / R module is close to the heat sink to achieve the purpose of rapid heat conduction.

2. The working principle and design scheme of the component

The highly integrated multi-channel tile-type T/R module proposed in this paper adopts bare chip to improve the system integration and reliability, makes full use of the space volume, realizes the interconnection of RF signal and low-frequency power supply control signal by vertical interconnection technology, and realizes high-density stacking of RF devices by LTCC technology. The block diagram of the module is shown in Figure. 1.

The T/R module integrates 4 transceiver channels, each of which is composed of transceiver branches. The received 4 channel signals are transferred to micro-strip line through SSMP1 connector, and then into limiter and LNA, then to multi-function module through conversion module 1, and then to power divider module through conversion module 2. After 4 channel synthesis, the signal is vertically transited to SSMP2 output.

The transmitted signal is transited from SSMP2 input to power divider module, divided into 4 channels, and then to multi-function module through conversion module 2, then to power amplifier module through conversion module 1, and finally transited to SSMP1 output to 4 transmission channels.

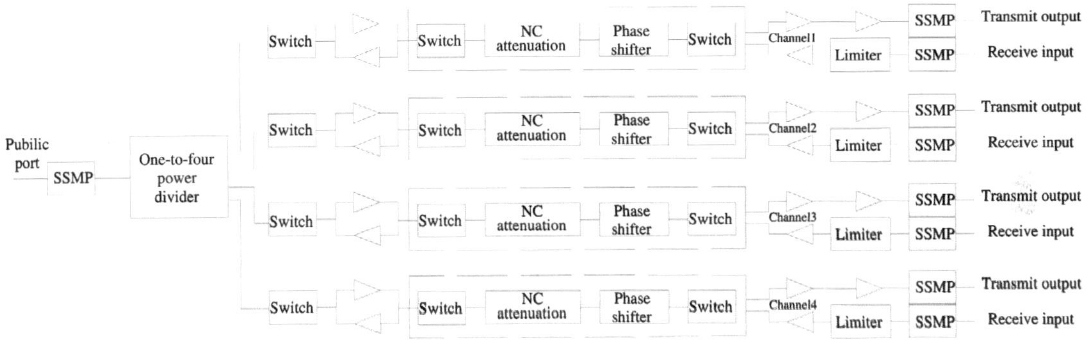

Figure 1. Module principle block diagram

The highly integrated multi-channel tile-type T/R module realizes three-layer integration by means of 3D stack. The main circuit is located in power amplifier module, multi-function module and power divider module. The section structure of T/R module is shown in Figure. 2.

The T/R module uses two LTCC multilayer to basically realize the high-density stacking of three-layer planar RF devices. Each layer is connected by the elastic button of the conversion module to realize the reliable connection between the RF signal and the control power signal. Each module is tested and tested separately by a special test fixture.

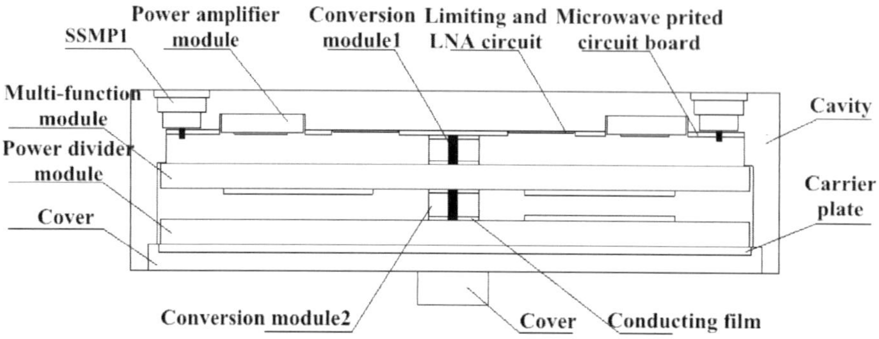

Figure 2. The section structure of T/R module

The first layer is power amplifier and LNA layer, which mainly includes power amplifier module, limiting and LNA circuit, SSMP connector 1, etc. it is installed on the aluminum alloy cavity to realize the functions of limiting amplitude and low noise amplifier of receiving channel and power amplification of transmitting channel. Among them, ong them, the mounting surface of the power amplifier module is in close contact with the heat sink, so that the heat generated by the power amplifier can be quickly everywhere, and the temperature of the power amplifier chip and the whole active antenna array can be reduced.

The second layer is the multi-functional module layer. The multi-functional module includes 4 channels, multi-functional chip and control chip, which are all bonded to the LTCC substrate with conductive adhesive. The gold wires are cascaded between the chip and the chip and between the chip and the substrate. The contacts on the front and back of the substrate are flexibly connected with conversion module 1 and conversion module 2, so as to realize the miniaturization and integration of the components.

The third layer is the power divider module layer. The power divider module consists of 4 channel driver chips, control chips and power supply chips. The front contact of the substrate is connected with the conversion module 2 flexibly, and the back contact is connected with SSMP2 and external connector. It mainly realizes the shunt synthesis of RF signal and power control signal.

Considering that the transmission path should be shielded and isolated to avoid signal interference, the space volume should be fully utilized in the structure, and the multi-layer circuit and ground circuit should be reasonably divided to solve the mutual interference problem of RF, power supply, digital signal wiring.

In addition, the conversion module, multi-function module and power divider module are placed in the aluminum alloy cavity, and the internal space of the cavity is divided into several closed small spaces, and the 4 transceiver channels are isolated from each other, which effectively solves the EMC problem between channels.

3. Simulation of key circuit design

Based on the above design scheme, a highly integrated multi-channel tile-type T/R module of X-band was designed with a size of 40mm×40mm×10mm.

3.1 Design of vertical interconnection based on LTCC

The T/R module is realized based on 3D stacked mode. The module is connected vertically and horizontally by customized connector to solve the problems of power supply, standing wave and power on the module.

The advantage of tile-type module is that the module is compressed in the longitudinal direction without increasing the area in the horizontal direction, which greatly reduces the volume and weight of the module, and solves the interconnection problem between the vertical and horizontal directions.

The carrier material of wool button(elastic button) is PTFE, and the material of wool button is cylinder beryllium copper wire. After the wool button carrier is installed into the wool button carrier, it is integrated into the metal partition wall.

The vertical interconnection design of the T/R module based on LTCC is mainly divided into 3 core parts.

In the first part, SSMP connector realizes 4 channel signal vertical transition. The SSMP connector is connected with the circulator externally, and connected with the microwave printed circuit board of the first layer in the vertical direction. The microwave printed circuit board is connected with the power amplifier module, limiting and LNA circuit with gold tape or gold wire, and then connected to the second layer multi-functional module through the conversion module 1.

Through the optimization design and Simulation of the SSMP connector to the multi-function module, as shown in the figure, the S11 of the vertical transmission structure is better than -15dB in the frequency band of 5~25GHz, which can be well applied to the 3D vertical transmission of the RF signal of the module.

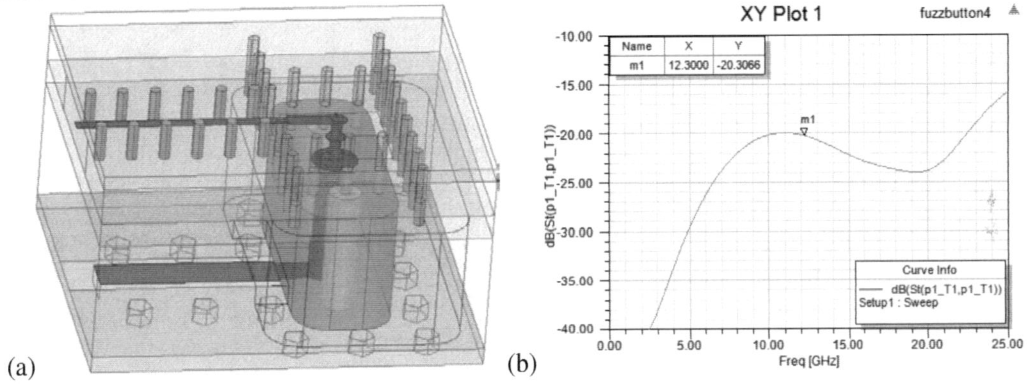

Figure 3. Simulation model and results of SSMP1 connector to multi-function module

In the second part, the front contact of the multi-functional module is connected to the front contact of the third layer power divider module through the conversion module 2. Through the optimization design and simulation of the multi-functional module to the power divider module, as shown in the figure, the S11 of the vertical transmission structure is better than - 10dB.

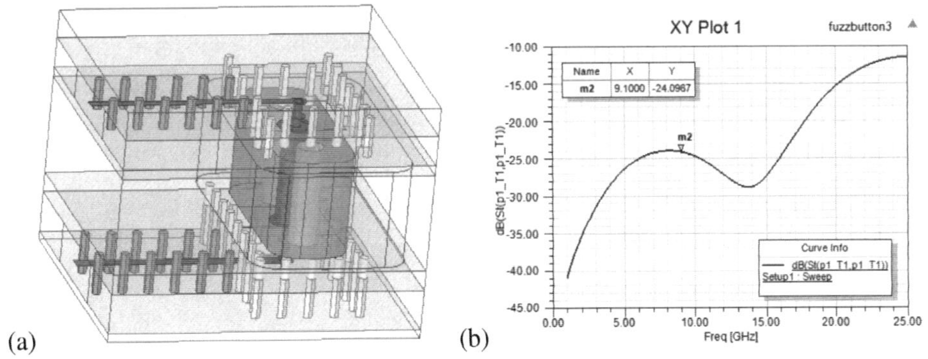

Figure 4. Simulation model and results of conversion module 2 of multi-function module to power division module

In the third part, the power divider module and carrier plate are sintered to the aluminum alloy cover plate with alloy solder, and the bottom contact of power divider module is elastically connected with the inner conductor of SSMP2 connector to output to the port. Through the optimization design and simulation of the power divider module to SSMP2 connector, as shown in the figure, the S11 of the vertical transmission structure is better than -20dB.

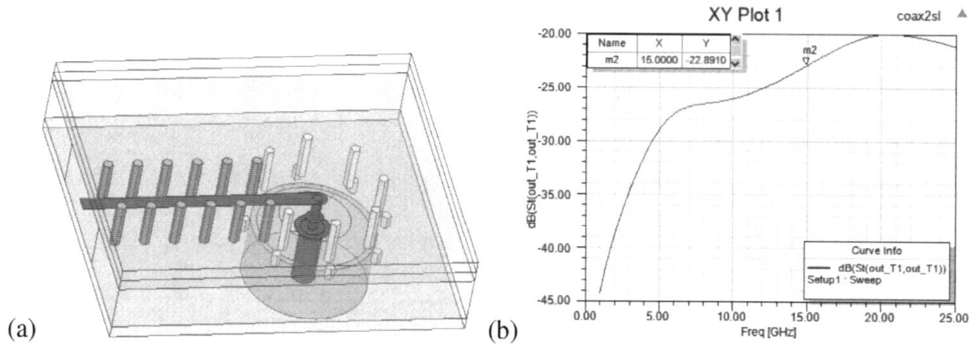

Figure 5. Simulation model and junction of power division module to SSMP2 connector

3.2 *One-to-four power divider*

The tile-type T/R module consists of 4 independent channels, which are isolated from each other. The RF signal and power control signal are synthesized by the third layer power divider module. Through the optimization design and Simulation of one-to-four power divider, as shown in the figure, all ports of the structure are better than -12dB, which meets the application requirements of components.

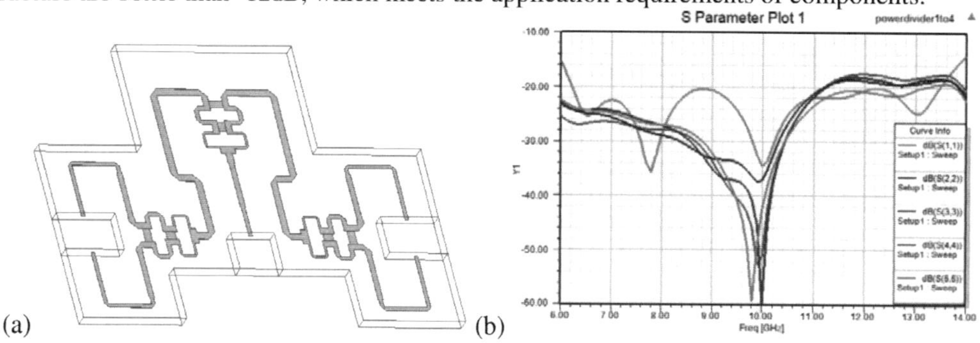

Figure 6. Simulation model and results of one-to-four power divider

3.3 *Transmission from LCTT flat layer microstrip line to stripline*

The coaxial-like structure is used between the flat layers of LTCC. The signal flows through the quasi coaxial-like in the micro-strip line and outputs to the stripline to realize the transition transmission in the flat layer, which fully guarantees the isolation of the transmission path. As shown in the figure, the transmission structure from micro-strip to stripline is optimized and simulated, and the ports of the structure are better than -22dB.

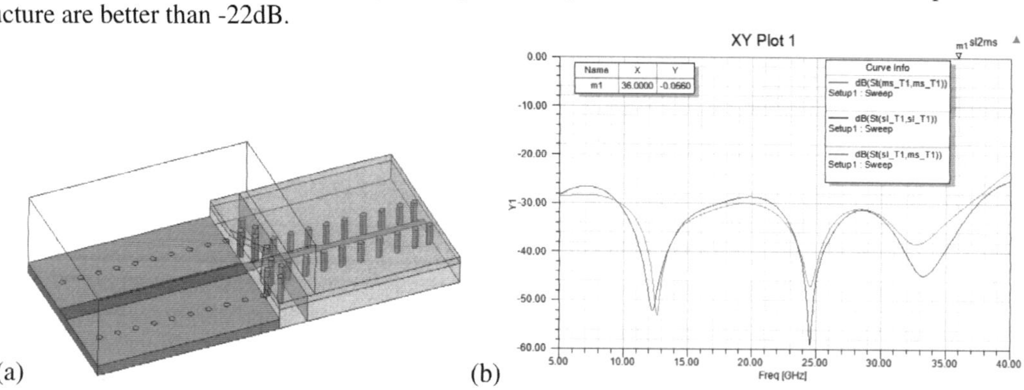

Figure 7. Simulation model and results of flat layer micro-strip to stripline

4. Concluding remarks

This paper presents a design method of vertical interconnection for highly integrated multi-channel tile-type T/R module. The module integrates 4 isolated transmission and reception channels with a size of 40mm×40mm×10mm. Compared with the traditional T/R module structure, it has the advantages of high reliability, miniaturization, lightweight and multi-functional integration. It can meet the application of active phased array antenna with limited space size, special loading platform or conformal array with low profile requirements.

5. References

[1] C. Lin, H. Liu and Y. Zheng, "Application of GaN high power chips in T/R modules," 2013 Proceedings of the International Symposium on Antennas & Propagation, Nanjing, China, 2013, pp:515-518.

[2] Anxue Zhang, Hao Hu, Jin Zhao, Zhuo Xu, Yansheng Jiang. "Design of Compact Filter Based on Multilayered Left-Handed Transmission Line Using LTCC Technology," Frequenz,2011,65:1-2.

[3] Fuzhou Tai, Jian Zhang. "Tile Type Multi-Channel Transceiver Module Applied for Phased Array Antenna. Journal of Computer and Communications, " 2018, 06(12):56-63.

[4] Rathod S, Sreenivasulu K, Beenamole K S. " Evolutionary Trends in Transmit/Receive Module for Active Phased Array Radars," Defence Science Journal, 2018, 68(6): 553-559.

[5] A. Di Carlofelice, F. de Paulis, A. Fina, U. "Di Marcantonio, A. Orlandi and P. Tognolatti. Compact and Reliable T/R Module Prototype for Advanced Space Active Electronically Steerable Antenna in 3-D LTCC Technology," IEEE Transactions on Microwave Theory and Techniques, 2018, 66(6): 2746-2756.

[6] Najam Muhammad Amin, Lianfeng Shen, Zhigong Wang, Ovais Akhter, "60GHz-Band Low-Noise Amplifier," Journal of Circuits Systems & Computers, 2016, 26(5): 1750075.

[7] Mark S. Hauhe , John J. Wooldridge, "High density packaging of X-band active array modules," IEEE Transactions on Components Packaging & Manufacturing Technology Part B, 2002, 20(3):279-291.

[8] Zhiguang Zhang , Xu Zheng, Xiao Liu, Xiangyang Qi,Zhe Xu, " A Vertical Interconnection Method in Tile T/R Module," ence Technology and Engineering, 2013, 13(11):3104-3108.

[9] Jinzhong Hao, Yu Zhang , Yang Zhou, "Design of a Broadband Multi-channel Tile-type T / R Module," Telecommunication Engineering, 2015 55(01):108-112.

Journal of Physics: Conference Series **2065** (2021) 012009 doi:10.1088/1742-6596/2065/1/012009

Development of GIS Switch State Judgment System Based on Image Recognition

WenHan Zhao[1], Feng Wen[2,*], Chen Han[2], Zhoujian Chu[1], Qingyue Yao[1], Kesong Ji[1]

[1] Maintenance Company, State Grid Jiangsu Electric Power Co.,Ltd., Nanjing, China
[2] School of Automation, Nanjing University of Science and Technology, Nanjing, China

Email: wen@njust.edu.cn

Abstract. Aiming at the fast opening and closing speed of the GIS isolation/grounding switch, manual observation is more difficult, so it is difficult to judge the current switch status. This paper proposes an OpenCV-based image identification algorithm to identify the position of the switch movable contact during the opening and closing process of the isolating switch, thereby judging the state of the isolating switch. This system uses Raspberry Pi as the main hardware core, the server drives the CMOS camera through Raspberry Pi 4B, collects image information in the GIS optical observation window, and performs simple processing, and transmits it to the Raspberry Pi 4B based on the UDP protocol as the main core. In the upper computer and adopt the target detection algorithm based on OpenCV to track the current isolation/grounding switch contact position and determine the current opening and closing state.

1. Introduction

GIS equipment is widely used in power systems due to its small footprint and stable performance. Since GIS equipment is a sealed equipment, compared with traditional open equipment, it is not intuitive, and it is impossible to directly observe the contact conditions of the contacts to determine whether the GIS isolation/grounding switch contacts are in place. Failure to close in place will cause equipment accidents, economic losses, and even endanger personal safety. In addition, when an accident occurs in a GIS equipment, it is difficult to deal with, the restoration of power supply is slow, and the cost of accident handling is high, which brings hidden dangers to the use of GIS equipment.

At present, some teams have conducted research on the GIS status determination device. [1-4]. Since the opening and closing of the GIS isolation/grounding switch and the mechanical state when there is a fault are different, the corresponding switch movable contact position is different, so the opening and closing state can be judged by observing the position of the movable contact. Image recognition technology includes three stages: image acquisition, target analysis and tracking, and state judgment. It is based on changes in visual effects brought about by changes in its mechanical state, tracking the movement state of the moving contact to determine the current switch's opening and closing state.

Compared with other methods, the use of image recognition technology to determine the state of the switch's moving contact is more intuitive [4-7]. However, the GIS equipment is a sealed equipment. Compared with the traditional AIS, it is not intuitive. It is not possible to directly observe the contact conditions to determine whether the GIS isolation/grounding switch contacts are in place. Therefore,

this article sets up a video sensor in the GIS optical observation window to collect the movement position of the moving contact inside the GIS in real time and uses the UDP protocol to transmit the collected video information and send it to the client. Image processing technology is used to determine the position of the current switch movable contact, and then determine the current switch state.
The structure of the remaining parts of this article is as follows. The second part introduces the structure and methods of the system, the third part describes the experimental structure and results, and the fourth part summarizes the full text and looks forward to the future work.

2. System design and proposed method

2.1. System Design and Moving Target Tracking Algorithm

Raspberry Pi is a palm-sized card computer developed by the British Raspberry Pi Foundation. As a palm-sized card computer, the Raspberry Pi has been widely used in hardware intelligence to create home theatres, wireless routers, FTP servers, face recognition, smart furniture, etc [7-8]. The light weight and high performance of Raspberry Pi provide a good solution in the application environment of this article.

The purpose of this system is to detect the opening and closing status of the GIS isolation/grounding switch. First, the image data is acquired by the CMOS sensor installed in the optical observation window of the GIS equipment (as shown in Fig 1), and the moving contact of the switch in operation showed in Fig 2. Then the data is transmitted to the terminal for real-time processing through transmission. The overall structure of the system is shown in Fig 3.

Since the GIS isolation/grounding switch has a fast action speed, the operation can be completed in an average of 2-3s. To improve the real-time performance and detection speed, this paper chooses OpenCV as the main core part of the image recognition algorithm. Since the background of the GIS isolation/grounding switch action area is relatively simple (as shown in Fig 2), considering the processing speed of the processor, this paper adopts the inter-frame phase difference method as the target tracking algorithm.

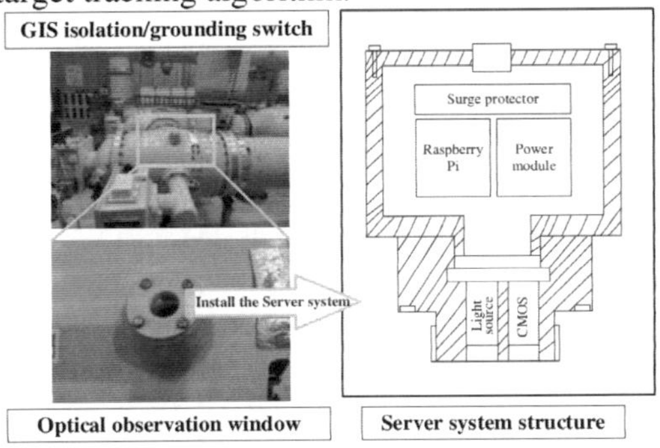

Figure 1. Optical Observation Window of GIS Equipment and Server System Structure.

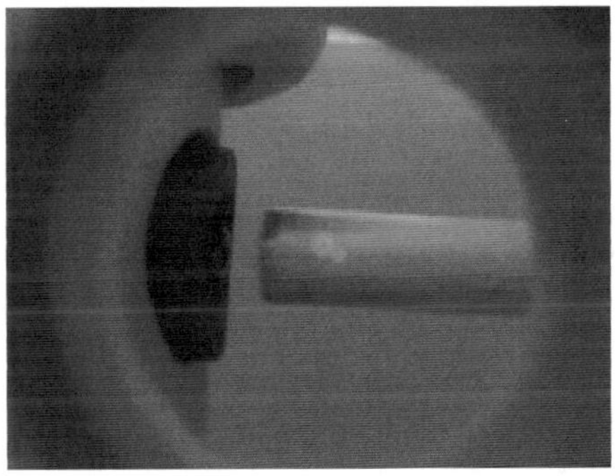

Figure 2. The moving contact of the switch in operation.

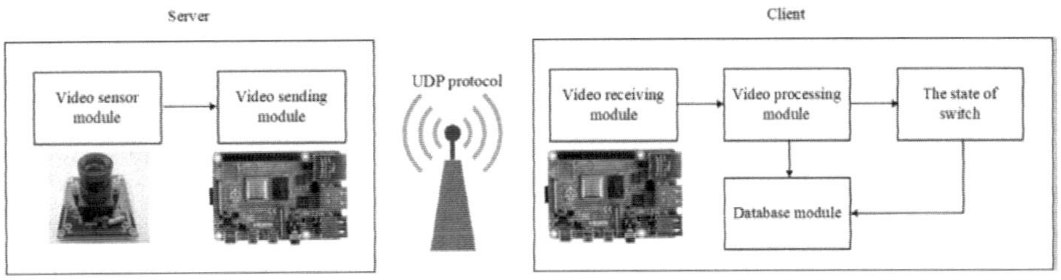

Figure 3. The overall structure of the system.

The inter-frame difference method is a method of marking moving objects by performing difference operations on two or three adjacent frames of images in the video. The advantages of the inter-frame difference method are that the algorithm is simple to implement, and the program design

complexity is low; it is not sensitive to scene changes such as light and can adapt to various dynamic environments with good stability. The disadvantage is that the complete area of the object cannot be extracted, there are "holes" inside the object, and only the boundary can be extracted.
The three-frame difference method is to take the "AND" operation in the two frame difference images of the adjacent three frames of pictures, namely:

$$D_n^{'}(x,y)=|F_{n+1}(x,y)-F_n(x,y)|\cap|F_n(x,y)-F_{n-1}(x,y)| \quad (1)$$

As shown in Fig 4,where F_{n+1} is the (n+1)th frame image, Fn is the (n)th frame image,F_{n-1} is the (n-1)th frame image, *x* and *y* are the pixel coordinates, *D* is the difference image, *D'* is the image after OR operation, then through threshold processing, connectivity analysis and discriminate,the current switch state can be determined by determining the relative distance between the current switch movable contact and the opening and closing position.

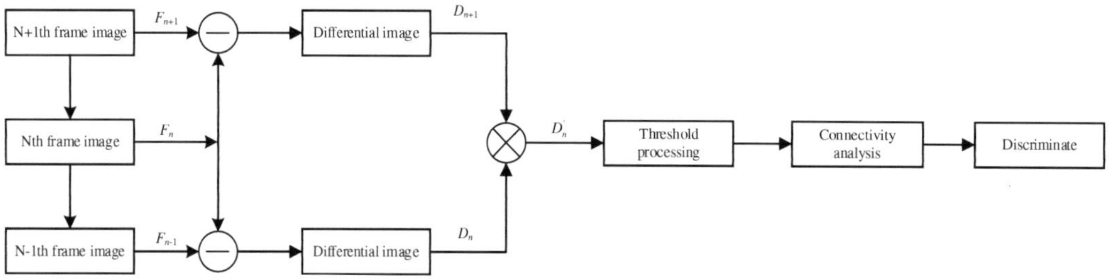

Figure 4. Three-frame difference algorithm flowchart.

2.2. Image Processing Algorithm Based on OpenCV
OpenCV is a cross-platform computer vision library based on the BSD license developed by Intel.
Firstly, pre-process the obtained image, convert the image into a grey image, and the histogram equalization is performed to obtain a new image with uniform grey distribution, and then median filtering is performed to achieve smoothing of the image.
Grayscale is a method of expressing image brightness, which is calculated by the following formula:

$$Y=\left(\frac{R^{2.2}+(1.5G)^{2.2}+(0.6B)^{2.2}}{1+1.5^{2.2}+0.6^{2.2}}\right)^{\frac{1}{2.2}} \quad (2)$$

Among them, Y represents the grey value of the image after grayscale, R represents the R channel component in the RGB image, G represents the G channel component in the GRB image, and B represents the B channel component in the RGB image.
Histogram equalization is to transform the histogram of the original image into a uniform histogram through a transformation function, and then modify the original image according to the uniform histogram to obtain a new image with uniform grey distribution. The equalization operation is defined by the following formula:

$$Z'=\frac{Z_{\max}}{S}\bullet\sum_{i=0}^{z}h(i) \quad (3)$$

Z' is the pixel value after equalization, *S* is the total number of pixels, Z_{max} is the maximum value of the pixel point, and *h(i)* represents the cumulative distribution function of the value *i*.
The median filter adopts a non-linear method, which is very effective in smoothing impulse noise. At the same time, it can protect the sharp edges of the image and select the appropriate point to replace the value of the polluted point. Therefore, the processing effect is good, and it is effective for the image sensor and the transmission channel. The black and white salt and pepper noise produced by decoding processing, etc. performs better. The size of the collected image is 640×480, and the 3×3 kernel is selected for filtering. The grey value at the centre of the kernel is:

$$Y(x,y)=median[Y(x\pm1,y\pm1),Y(x,y\pm1),Y(x\pm1,y),Y(x,y)] \quad (4)$$

After the pre-processing, using the three-frame difference method to detect the moving target, determine the state of the current switch movable contact.

2.3. GIS Switch State Judgment Algorithm

Combining OpenCV and the three-frame difference method, the judgment process of the GIS switch state based on the Raspberry Pi platform is shown in the Fig 5.

When preparing for opening and closing operations, the server and client are initialized, ready to transmit and receive data. The client acquires the transmission image, and then pre-processes the image based on OpenCV, then uses the three-frame difference method to track the movement of the switch moving contact, and finally determines the switch state according to the current position and movement direction of the current moving contact.

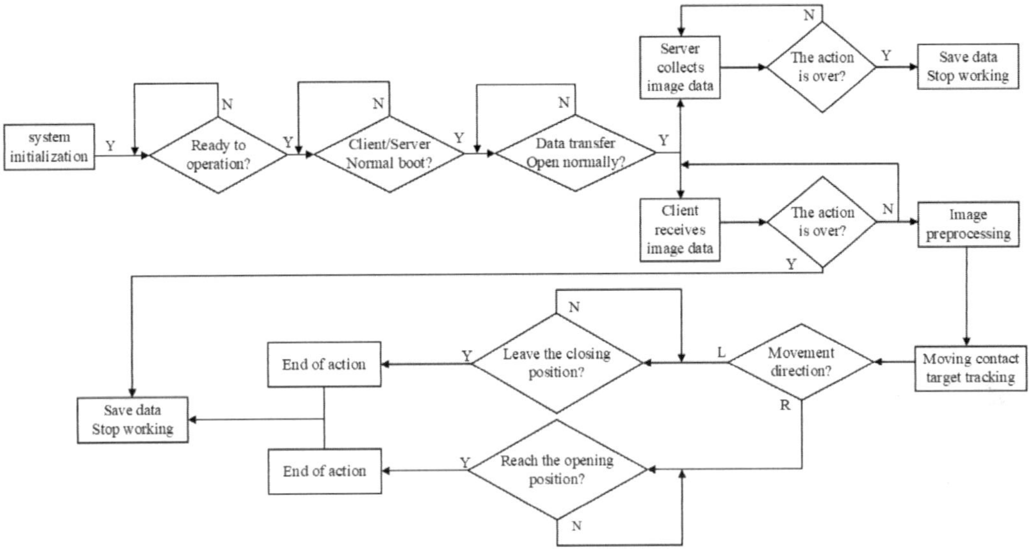

Figure 5. GIS switch state judgment process.

3. Experimental result

The experimental platform is shown in Fig. 6. In the early stage, we recorded a GIS isolation/grounding switch action video through a CMOS sensor to verify the performance of the proposed system.

The algorithm is tested on the Raspberry Pi 4B+ platform. The video processing and target recognition speed can reach about 29FPS (as shown in Fig .7). Although it is found that the outer contour of the identified object is slightly larger than the area of the moving contact, compared to the size of the moving contact itself, the contour error is small and will not interfere with the operation of the system to determine the position of the moving contact.

Figure 6. Experiment platform.

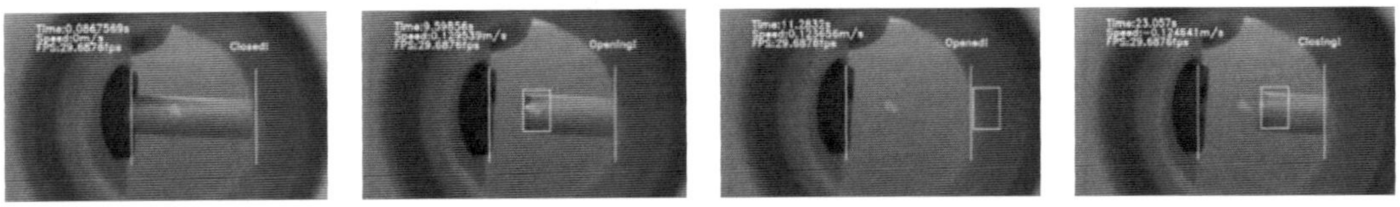

Figure 7. Experiment result.

4. Conclusion

This paper proposes an OpenCV-based image recognition algorithm to identify the position of the switch movable contact during the opening and closing of the isolating switch, thereby judging the state of the isolating switch. This system uses Raspberry Pi as the main hardware core, the server drives the CMOS camera through Raspberry Pi 4B, collects image information in the GIS optical observation window, and performs simple processing, and transmits it to the Raspberry Pi 4B as the main core based on UDP protocol. In the host computer, and use the OpenCV-based three-frame difference algorithm to track the current isolation/grounding switch contact position, and achieve a detection speed of about 20FPS, it basically meets the requirements of real-time detection of switch opening and closing status. Later, the image recognition algorithm will be further optimized to reduce the error of the outer contour of the object, to determine the current switch movable contact position more accurately.

Acknowledgement

This work was Funded by China Postdoctoral Science Foundation (2020M671498), the Basic Research Program of Jiangsu Province (BK20180485), the Fundamental Research Funds for the Central Universities (30919011241), the Jiangsu Planned Projects for Postdoctoral Research Funds (2020Z374), and the Science and Technology Project of State Grid Jiangsu Electric Power Co.,Ltd. (J2020135, SGJSJX00BGJS2002373).

5. References

[1] Wu Ruxiang. Application research of GIS disconnecting switch operating characteristics state monitoring [D]. *South China University of Technology*, 2015.

[2] U SCHICHLER, W KOLTUNOWICZ, F ENDO, et al.Risk Assessment on Defects in GIS Based on PD Diagnostics. *IEEE Transactions on Dielectrics and Electrical Insulation*, 2013, **20(6):**2165-2172.

[3] T.HOSHINO , K.KATO , NAOKI HAYAKAWA et al. Frequency Characteristics of Electromagnetic Wave Radiated From GIS Apertures. *IEEE Transactions on Power Delivery*, 2001,**16(4)**552-557.

[4] Yulong Huang. Research on the Motor Drive Based Operating Mechanism for High Voltage Circuit Breaker. *Proceedings of International Industrial Informatics and Computer Engineering Conference*, 2015.

[5] Mao Wenqi, Wang Bozhong, Jiang Yizhou, Yuan Pei, Huang Mingwei, Zhou Hui, Zhou Bin, Wei Juan. Summary and Prospect of Breaking-closing Position Monitorring Method for High Voltage Disconnector. [J]. *Smart Power*, 2019,**47(08)**:112-119.

[6] Fang Sheng, Shu Xiaohua, and Li Dewu. A Method Based on Machine Vision for Opening-Closing Status Recognition of Substation Disconnecting Switches. [J] *Journal of Hunan University of Technology* 2017,**31(06)**:32-36.

[7] Victor Gonzalez-Huitron, José A. León-Borges, A.E. Rodriguez-Mata, Leonel Ernesto Amabilis-Sosa, Blenda Ramírez-Pereda, Hector Rodriguez, Disease detection in tomato leaves via CNN with lightweight architectures implemented in Raspberry Pi 4, *Computers and Electronics in Agriculture*, Volume **181**, 2021, **105951**, ISSN 0168-1699.

[8] D. K. Dewangan and S. P. Sahu, Deep Learning-Based Speed Bump Detection Model for Intelligent Vehicle System Using Raspberry Pi, *IEEE Sensors Journal*, vol. **21**, no. **3**, pp. 3570-3578, 1 Feb.1, 2021, doi: 10.1109/JSEN.2020.3027097.

Journal of Physics: Conference Series **2065** (2021) 012010 doi:10.1088/1742-6596/2065/1/012010

LFT Bond Graph for Online Robust Fault Detection and Isolation of Hybrid Multi-Source System

Mahdi Boukerdja[1,2*], Youness Radi[1], Omprakash[1], Sumit Sood[1], Belkacem Ould-Bouamama[1], Aissa Chouder[2], Anne-Lise Gehin[1], Jean-Yves Dieulot[1] and Mathieu Bressel[1].

[1]CRISTAL UMR CNRS 9189. Université de Lille. 59655 Villeneuve d'Ascq. France.
[2]University of Msila. Electrical Engineering Laboratory (LGE). 28000 BP166 Msila. Algeria.

Mahdi.boukerdja@univ-lille.fr,mahdi.boukerdja@univ-msila.dz

Abstract. Green hydrogen is undoubtedly the most promising energy vector of the future because it is captured by renewable and inexhaustible sources, such as wind and/or solar energy, and can be stored over the long in high-pressure cylinders, which can be used to feed the fuel cells to produce the electricity without emitting any pollutants. The system incorporated renewable sources and process used to produce the green hydrogen is the hybrid multi-source system (HMS). The production of hydrogen needs a reliable HMS, which always requires online monitoring for real-time Fault Detection and Isolation (FDI) because the risk of accidents in HMS and safety issues increases due to the possibility of faults. However, online monitoring of FDI is challenging due to the multi-physics dynamics of HMS and the inclusion of uncertain parameters and several disturbances. This paper proposes an online robust fault detection algorithm to detect system faults based on the properties of the graphical linear fractional transformation bond graph (LFT-BG) modeling approach. Here, the analytical redundancy relations (ARRs) and their uncertain parts extracted from the LFT-BG model are used to develop an online robust FDI algorithm for HMS. Numerical evaluations of ARRs and their uncertain parts, respectively, generate the residual signals known as "faults indicators" and their uncertain bounds known as "adaptive thresholds." These thresholds evolve with system variables in the presence of parameter uncertainties for ensuring robust FDI for HMS to minimize false alarms. The validation of this approach is carried out using 20sim software that is familiar with BG modeling.

1.Introduction

Due to the environmental protection requirements, the Hybrid multi-source system (HMS) is involved in the task of storing renewable energy in the form of hydrogen gaze fuel [1]. It might have the same configurations as illustrated in figure 1, which consists of a parallel interconnection of wind turbine (WT) and photovoltaic (PV) generators. The flow of power produced by the WT and PV generators is controlled by AC/DC and DC/DC converters, respectively, to feed the Hydrogen production system (HPS), which usually requires constant power consumption to realize the Hydrogen production task. For that, the storage device (SD) is associated with HMS elements to ensure the constant power consumption of HPS. The SD illustrated in figure 1 consists of a lead-acid battery and DC/DC bidirectional converter. The DC/DC bidirectional converter of the SD controls the battery power flow in the opposite and direct direction, which allows storing the power surplus produced by WT and PV, and compensating for the power deficiency when WT and PV do not ensure the needed power of HPS.

The production of hydrogen fuel needs a reliable HMS, which requires online monitoring for real-time fault detection and Isolation (FDI) because the risk of accidents in HMS and safety issues increase due to the possibility of faults. However, online monitoring for FDI is challenging due to the multi-physics, uncertain dynamics of HMS and the inclusion of uncertainties of the measurements and parameters, and disturbances. These uncertainties complicate the task of FDI in HMS and cause false alarms [2].

The model-based techniques (such as parity space and observers-based techniques) can deal with the FDI issues [3, 4]. The parity space model-based method is intended only for the linear systems model. This method derives the analytical redundancy relations (ARRs), which are given in terms of known signals, variables, and parameters in closed mathematical forms, from the differential equations of the system dynamics presented in the state-space form [3]. On the other hand, the systems model can be used to design the observers for FDI, where the observer's estimation errors represent the fault indicators.

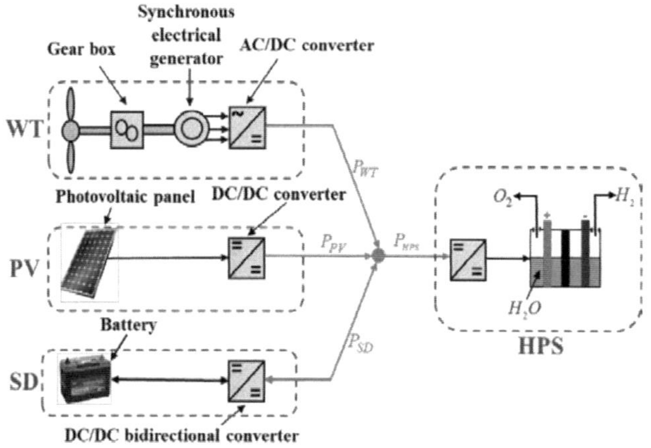

Figure 1. HMS configuration.

The observer technique can also be applied for the model of the nonlinear system [4]. The residual signals derived from both methods are sensitive to faults, and they are limited to the FDI of sensors and actuators but not of the parametric (component) faults. Thus, the parametric faults corresponding to the HMS component cannot be detected under these methods. Because of many components of the HMS, simplistic models would yield no robust and poor FDI performance.

This paper addresses the FDI model-based issues by exploiting the graphical linear fractional transformation bond graph (LFT-BG) to design online monitoring of HMS components for FDI. As the LFT-BG modeling approach models all HMS physics and includes disturbances, process, parameter, and measurement uncertainties, FDI's proposed method earns the robustness against all uncertainties to prevent false alarms. The LFT-BG modeling approach is known by the causal properties (derivative and integral causal properties) representing the cause and effect between system variables. These causal properties help to generate directly from the uncertain bond graph model the fault indicators by decoupling the uncertainties from the faults and perform the isolability and monitorability analysis based on fault signature matrix (FSM), which is exploited to verify the ability of the proposed approach to localize the fault or set of faults in the HMS [6]. The derivative causal property is preferred for FDI to avoid unknown initial conditions. While the integral causal property is preferred for behavioral and simulation studies [5]. The HMS LFT-BG model developed under the derivative causal property is used to derive *ARRs* and their uncertain parts in terms of known signals, variables, and parameters in closed mathematical forms. Numerical evaluations of *ARRs* and their uncertain parts generate the residual signals that are "fault indicators" and their uncertain bounds that are "adaptive thresholds," respectively. These thresholds evolve with system variables in the presence of uncertainties [5,9]. The proposed approach monitors the WT, PV, and SD components without including the power converters in the study and validated by using 20sim software.

The paper contains four other sections. Section 2 presents a brief theory of LFT-BG to derive the online robust monitoring for FDI. Section 3 presents the process to develop online robust monitoring for FDI and monitorability and isolability analysis for fault localization using the HMS's LFT BG model. Section 4 shows the simulation study, and this paper is ended with a conclusion.

2. Bond graph theory for fault detection and isolation

2.1 Conventional bond graph theory

BG is a graphical language used to model multi-physical systems. It is a union of two sets of elements denoted by $S = \theta \cup A$, where A is a set of power bonds denoted by the half arrows, which show the exchange of power that is a product of effort e and flow f between the system components. In contrast, θ is a set of elements used to model the system components. The elements of θ are the resistance R, capacitance C, inductance I, transformer TF, gyrator GY, effort source Se, flow source Sf, effort sensor De and flow sensor Df. The element R models the system components having the linear dynamic. The elements C and I model the system components having the same dynamics as the capacitance and inductance, respectively, which correspond to the system components storing the energy. The elements Se and Sf model the system components generating the power in the form of effort and flow, respectively. The elements TF and GY model the system components transforming the power without causing any loss. The elements De and Df represent the sensors that obtain the effort and flow measurements, respectively. The whole BG elements are connected through the power bonds and junctions "1" and "0" by respecting their properties (flow equality and effort equality, respectively) [6].

Moreover, BG formalism is known by the causal properties, which are the integral and derivative causal properties. These properties are the way to know how the system components are acting between them, as illustrated in figure 2. The causal properties are presented in BG by the cross-stroke in one of the power bond sides. S_1 acts on S_2 by an effort, and S_2 replies to S_1 by a flow (effort causality) if the cross-stroke is placed on the right side. On the other hand, the cross-stroke placement on the left side shows S_1 acts on S_2 by a flow, and S_2 replies to S_1 by an effort (flow causality). The use of integral and derivative causal properties is to simulate the dynamic of modeled multi-physical systems and develop online monitoring for FDI, respectively, before the online implementation [9]. The cross-stroke assignment in integral and derivative causal properties is performed by respecting the rules, which are well explained in [6,8]. Under derivative causal property, the elements of effort and flow sensors De and Df become the effort and flow signal sensors SSe and SSf, respectively [8].

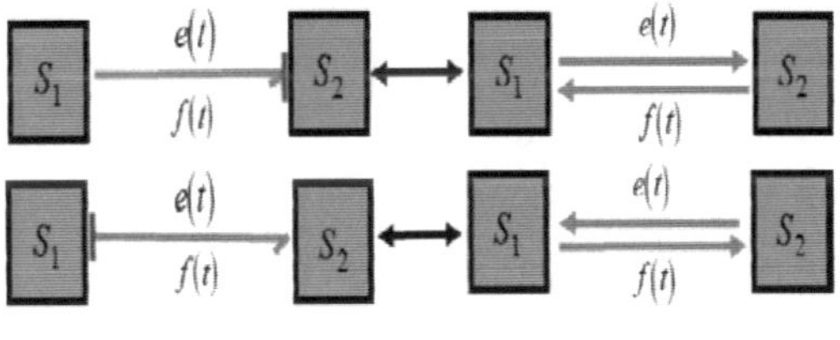

Figure 2. Causality presentation.

2.2 Linear fractional transformation for modeling the uncertainties in bond graph

The HMS parameter identification bears difficulties, which prevents obtaining the nominal values of HMS parameters in the real process. In this case, each HMS parameter can be expressed by an interval of values, where its minimum and maximum values are related to the parameter's uncertainty. Also, The HMS parameters may vary in the range of values, which can also be related to the parameter uncertainties, due to unknown disturbance in the absence of faults. This issue complicates FDI's task for HMS and is considered as the main cause of false alarms. For more robustness of the FDI algorithm, LFT presentation is used to incorporate the uncertain part of the HMS parameter in the BG model [5,9].

Using the LFT methodology, uncertainties on parameters are considered as an additive or multiplicative, as written in (1) and (2) where β is an example of any system parameter to illustrate the

LFT methodology, β_n is the β nominal value, δ_β is the relative uncertainty, and Δ_β is the absolute uncertainty.

$$\beta = \beta_n \pm \Delta\beta \quad (1)$$

$$\beta = \beta_n\left(1 \pm \delta_\beta\right) \quad (2)$$

To illustrate how uncertainty is integrated into the BG using the LFT presentation methodology, consider the BG R element's case. Let consider $R=R_n(1\pm\delta_R)$ the real value of the R element.

Assume that the causality of this element R is assigned in flow causality. Then, the effort of R is written as follows:

$$\begin{aligned} e_R &= R\cdot f_R = R_n\left(1\pm\delta_R\right)f_R = R_n f_R \pm \delta_R R_n f_R \\ &= e_{R_n} + e_{\delta_R} \end{aligned} \quad (3)$$

According to (3), the uncertainty of R integrated into BG under flow causality is given as shown in figure 3a, where De^* is used to measure e_{Rn}, and $e_{\delta R}$ is introduced by MSe^*.

Assume now the causality of R is assigned in effort causality. Then, the flow is given as follows:

$$\begin{aligned} f_R &= \frac{1}{R}e_R = \frac{e_R}{R_n\left(1\pm\delta_R\right)} = \frac{1}{R_n}\left(1\pm\delta_{1/R}\right)e_R \\ &= \frac{e_R}{R_n} \pm \frac{\delta_{1/R}e_R}{R_n} = f_{R_n} + f_{\delta_{1/R}} \end{aligned} \quad (4)$$

according to (4), the uncertainty of R integrated into BG under effort causality is given as shown in figure 3b, where Df^* is used to measure f_{Rn}, and MSf^* introduces $f_{\delta 1/R}$.

The LFT presentation of elements C, I, GY, and TF in BG can be performed in the same way as the element R [9].

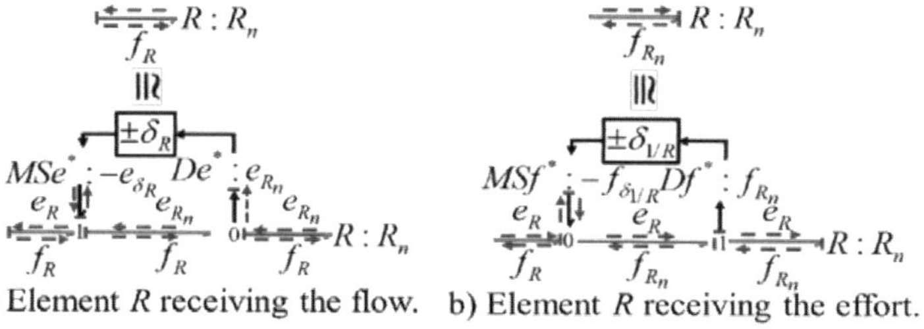

Figure 3. Causality presentation.

2.3 Online robust fault detection based on the linear fractional transformation bond graph

The design process of the online robust FDI consists of the ARRs, and their adaptive thresholds, which are developed by the use of the properties of the junctions ("0" and "1") where there is a SSe or SSf that correspond to the sensors used for monitoring [7,9]. The ARR is the sum of incoming power bonds efforts minus the sum of outgoing power bonds efforts if junction "1" is used. Also, the ARR can be the sum of incoming power bonds flows minus the sum of outgoing power bonds flows if the junction "0" is used. All ARR parameters must be known, and the uncertain parts are neglected [7]. The mathematical expression of ARR can be written as follows:

$$\begin{aligned} ARR &= \sum\left(e_{in_n} + \cancel{e_{in\delta}}^{0}\right) - \sum\left(e_{out_n} + \cancel{e_{out\delta}}^{0}\right) \\ &= \Psi\left(\sum SSe, \sum SSf, \sum Se, \sum Sf, \theta_n\right) \end{aligned} \quad (5)$$

or

$$\begin{aligned} ARR &= \sum\left(f_{in_n} + \cancel{f_{in\delta}}^{0}\right) - \sum\left(f_{out_n} + \cancel{f_{out\delta}}^{0}\right) \\ &= \Psi\left(\sum SSe, \sum SSf, \sum Se, \sum Sf, \theta_n\right) \end{aligned} \quad (6)$$

where e_{in}, e_{out}, f_{in}, f_{out}, Θ_n are the nominal incoming effort, outgoing effort, incoming flow, outgoing flow, and BG elements parameters, and $e_{in\delta}$, $e_{out\delta}$, $f_{in\delta}$, $f_{out\delta}$ are the neglected incoming effort, outgoing effort, incoming flow, and outgoing flow, respectively, generated by the uncertain parts. The adaptive threshold a is the sum of the absolute value of uncertain parts neglected in (5) or (6). It can be written as follows:

$$a = \sum\left|e_{in_\delta}\right| + \sum\left|e_{out_\delta}\right| = \Psi\left(\sum SSe, \sum SSf, \sum Sf, \sum Se, \theta_n, \delta_\theta\right) \quad (7)$$

or

$$a = \sum\left|f_{in_\delta}\right| + \sum\left|f_{out_\delta}\right| = \Psi\left(\sum SSe, \sum SSf, \sum Sf, \sum Se, \theta_n, \delta_\theta\right) \quad (8)$$

where $\delta_\Theta=\{\delta_R,\delta_C,\delta_I,\delta_{TF},\delta_{TF}\}$ and $\Theta_n=\{R_n,C_n,I_n,GY_n,TF_n\}$. The fault indicators are the numerical evaluation of *ARRs* called "residual signals." $r=eval(ARR)$. Theoretically, r should be zero, but r is not zero due to noises and parameter uncertainties. If $-|a|\leq r\leq|a|$, then the system operates in the nominal conditions (no faults). Else, faults occur in the system [5].

3.Online robust fault detection and isolation for hybrid multi-source system

In this section, the process to obtain the LFT-BG model of the HMS, which has the same configuration presented in section 1 (see figure 1), under the derivative causal property is briefly explained. It is obtained by joining the nominal BG models of HMS components in the single model and including the parameter uncertainties using LFT presentation. The obtained HMS LFT BG model will be used to drive the *ARRs* and their adaptive thresholds to monitor the WT, PV, and SD in the presence of uncertainties without considering the power converters.

WT component consists of a mechanical part deriving the synchronous electrical generator. The mechanical part is illustrated in figure 4a, and its BG nominal model is illustrated in figure 4b. It consists of the propeller driving a gearbox. This mechanical part is connected to the electrical synchronous machine through a shaft characterized by two parameters: J is the inertia moment, and σ is the viscous friction coefficient. The propeller, which is the generator of torque *Tear* deduced from the wind speed W_S and direction W_D, converts the wind's kinetic power to mechanical power. The Gearbox provides the suitable torque and rotational speed Ω to derive the synchronous electrical generator for generating the electrical power, where the torque provided by Gearbox is equal to $Tear/G=Jd\Omega/dt+f\Omega$ [10]. In the BG approach, the power variables are as follows: e is the torque, and f is the rotational speed. The source of effort *MSe* models the propeller: *Tear* and the Gearbox by *TF:1/G* element. As the same Ω drives all shaft particles, the shaft part is modeled by *I: J* and *R:σ* connected to the junction "1".

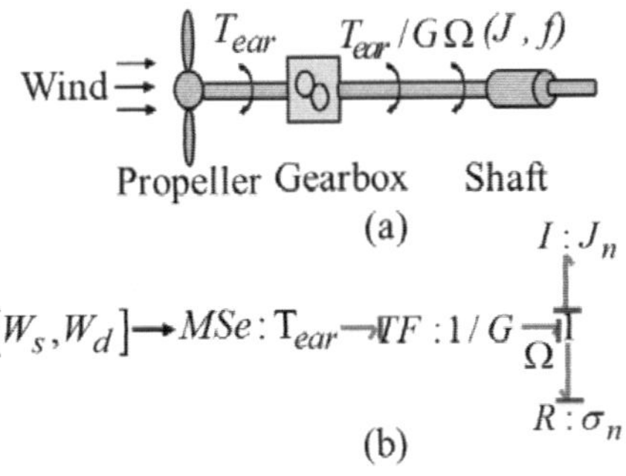

Figure 4. a) diagram of WT mechanical part, b) nominal BG model of WT mechanical part.

The synchronous electrical generator of the WT component illustrated in figure 5a consists of two parts: the moving and the fixed parts. It converts the mechanical power provided by the mechanical part to the electrical power when the moving part rotates due to shaft rotation. The permanent magnet used to make the moving part generates a magnetic field circulation, which excites the fixed part's winding to produce the electrical current. The nominal BG model illustrated in figure 5b of the synchronous electrical generator (power variables: e is the voltage; f is current) is obtained based on the park transformation technique [11]. The elements $MGY{:}3/2pl_{dn}i_q\Omega$, $MGY{:}-3/2pl_{dn}i_d\Omega$ and $GY{:}3/2p\lambda\Omega$ model the interaction of magnetic field circulation with the winding of the fixed part when the moving part is rotating, p is the generator pole pairs, and λ is the flux density of the permanent magnet, and the elements $R{:}\ R_{dn}$, $R{:}\ R_{qn}$, $I{:}\ L_{dn}$, $I{:}\ L_{qn}$, v_d, v_q, i_d and i_q refer to the electrical characteristic of the fixed part winding, where R_{dn} and R_{qn} are the resistances in the direct and quadratic axis, respectively, L_{dn} and L_{qn} are the inductances in direct and quadratic axis, respectively, v_d and v_q are the winding voltages in the direct and quadratic axis, respectively, and i_d and i_q are the winding currents in the direct and quadratic axis, respectively. The synchronous electrical generator's real outputs are i_a, i_b, i_c, v_a, v_b, and v_c. They are obtained using the inverse Park transformation of v_d, v_q, i_d and i_q.

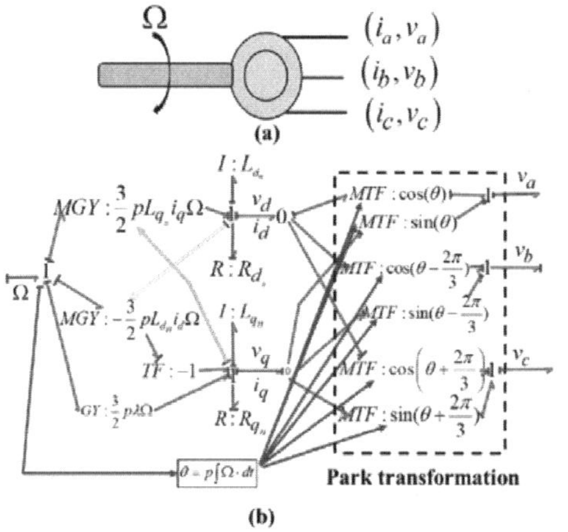

Figure 5. a) diagram of the WT synchronous electrical generator, b) nominal BG model of the WT synchronous electrical generator.

PV Component consists of a series of solar cells that capture solar irradiation (S) and convert it to electrical power. This PV component can be considered as the current source producing i_{ph} as depicted in figure 6a, and its nominal BG model is illustrated in figure 6b, where $MSf{:}i_{ph}$ is used to model the i_{ph} current source and $R{:}\ R_{shn}$, $R{:}\ R_{Dn}$, and $R{:}\ R_{sn}$ are used to model the particles causing the power loss, where R_{shn}, R_{Dn}, and R_{sn} are the equivalent shunt resistance, the equivalent resistance of diode deduced from the semi-conductor characteristic of the panel, and equivalent series resistance, respectively. This component's output power is the product of i_p and v_p, which are photovoltaic current and voltage, respectively [12].

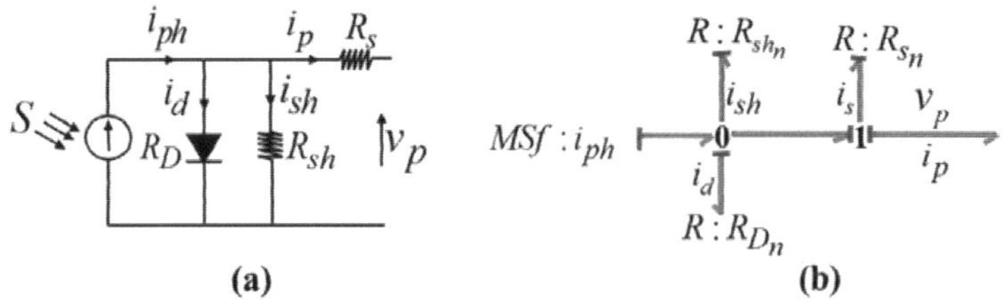

Figure 6. a) Diagram of the PV panel equivalent circuit, b) nominal BG model of PV.

The SD component consists of the lead-acid battery, which can be considered as a voltage source V_{oc} (the V_{oc} value depends on the state of charge *SOC* of battery) and represented by the equivalent circuit as shown in figure 7a. The nominal BG model of this component is illustrated in figure 7b, where *MSe:* V_{oc} is used to model the voltage sources, *C:* C_{dl} is used to model the double-layer capacitance of lead-acid battery, and *R:* R_{dl} and *R:* R_i are used to model the particles causing the loss of power, which are the charge transfer and internal resistances, respectively. This component's output power is the product of i_b and v_b, which are the lead-acid battery current and voltage, respectively.

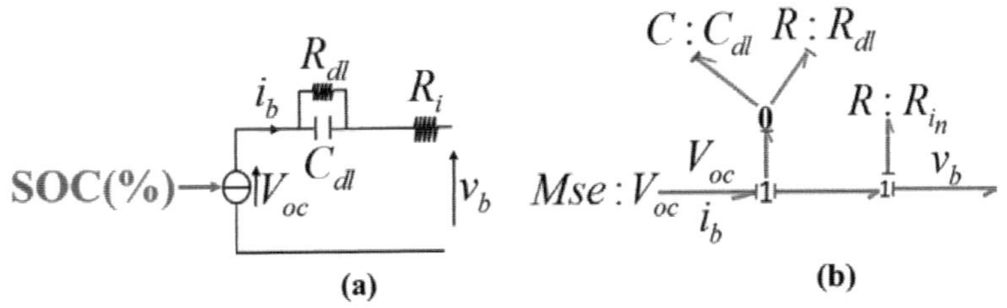

Figure 7. a) Equivalent circuit of lead-acid battery, b) Nominal BG model of SD component.

Regrouping the nominal BG models of the HMS components with including the uncertainties of the parameters using the LFT presentation, make the HMS LFT-BG model, as illustrated in figure 8. The sensors used to monitor the HMS components are SSf_1, SSf_2, SSf_3, SS_{f4}, SSf_5, and SSe_1, which measure Ω_m, i_{dm}, i_{qm}, i_{pm}, i_{bm}, *and* v_{cdlm}, respectively. In fact, there are no real sensors devices to measure i_{dm}, i_{qm}, and v_{cdlm}. These measurements are performed by inverse Park transformation (measuring first i_a, i_b, and i_c, and then using park transformation to deduce i_{dm} and i_{qm}) and observer technique, respectively [11,4].

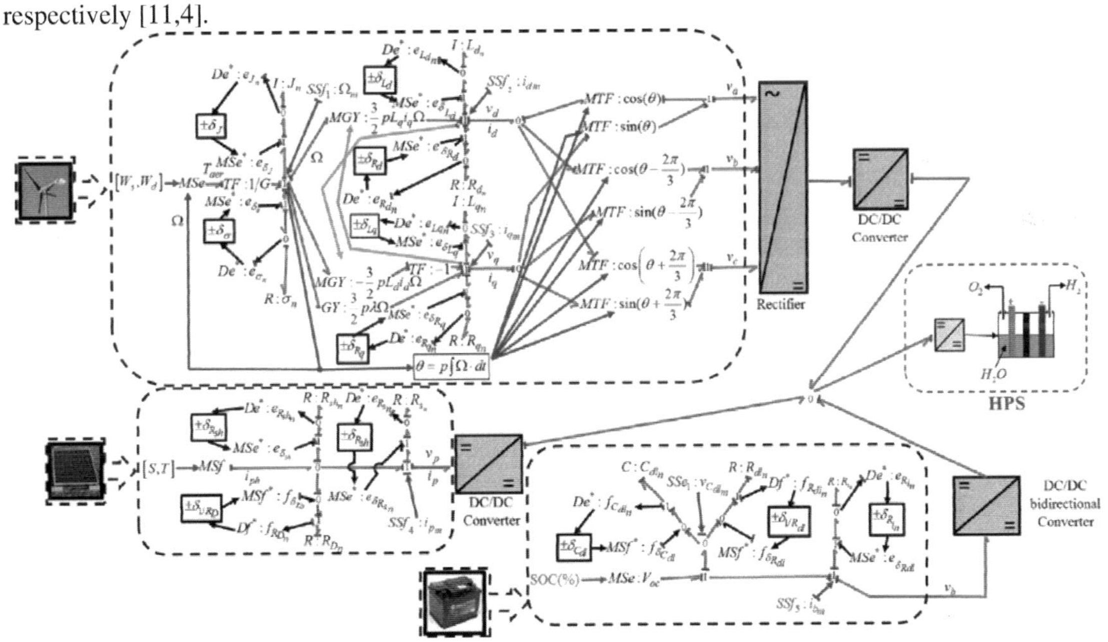

Figure 8. HMS LFT-BG model under derivative causality property.

Figure 9 presents a part (corresponds to the SD component) of the HMS LFT-BG model to illustrate the process used to derive the *ARRs* and adaptive thresholds using the definition presented in section 2.3. The junction used to drive the *ARR* is the junction where the monitoring sensor SSe_1:v_{cdlm} is placed.

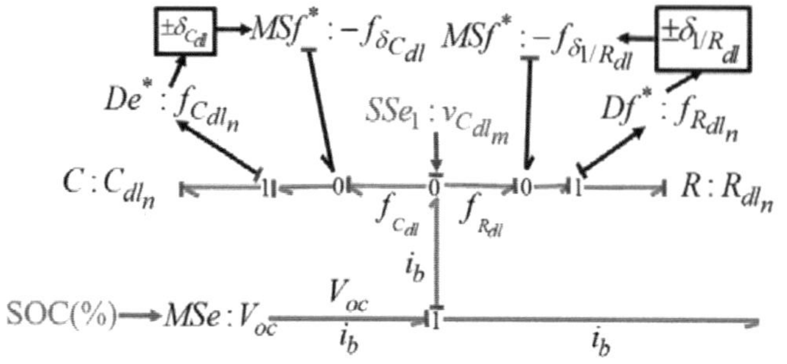

Figure 9. Part of the HMS LFT-BG model for illustrating the design process of *ARRs* and their adaptive thresholds.

It is obtained that :

$$i_b - f_{C_{dl}} - f_{R_{dl}} = i_b - \left(f_{C_{dl_n}} + f_{\delta C_{dl}}\right) - \left(f_{R_{dl_n}} + f_{\delta 1/R_{dl}}\right) \tag{9}$$

Neglecting $f_{\delta Cdl}$ and $f_{\delta 1/Rdl}$, the mathematical expression of the resulted *ARR* is written as follow:

$$ARR_5 = i_b - f_{C_{dl_n}} - f_{R_{dl_n}} = i_{b_m} - C_{dl_n}\frac{dv_{C_{dl_m}}}{dt} - \frac{v_{C_{dl_m}}}{R_{dl_n}} \tag{10}$$

where i_{bm} is the measurement value of i_b obtained by SSf_5. The adaptive threshold associated with *ARR* is the absolute sum of $f_{\delta Cdl}$ and $f_{\delta 1/Rdl}$, which can be written as follows:

$$0\,a_5 = \left|f_{\delta C_{dl}}\right| + \left|f_{\delta 1/R}\right| = \left|\pm\delta_{C_{dl}} C_{dl_n}\frac{dv_{C_{dl_m}}}{dt}\right| + \left|\pm\delta_{1/R_{dl}}\frac{v_{C_{dl_m}}}{R_{dl_n}}\right| \tag{11}$$

Then, other *ARR*s and their adaptive thresholds, which are used to monitor the HMS components, will be developed as explained in the previous example. They can be written as follows:

$$ARR_1 = \frac{3}{2} pL_{q_n} i_{q_m}\Omega_m - L_{d_n}\frac{di_{d_m}}{dt} - R_{d_n} i_{d_m} - v_{d_m} \tag{12}$$

$$ARR_2 = \frac{3}{2} p\Omega_m\left(\lambda - L_{d_n} i_{d_m}\right) - L_{q_n}\frac{di_{q_m}}{dt} - R_{q_n} i_{q_m} - v_{q_m} \tag{13}$$

$$ARR_3 = T_{ear} - J_n\frac{d\Omega_m}{dt} - \sigma_n\Omega_m - \frac{3}{2} p\lambda i_{q_m} \tag{14}$$

$$ARR_4 = \frac{R_{sh_n} R_{D_n}}{R_{sh_n} + R_{D_n}}\left(i_{ph} - i_{p_m}\right) - R_{s_n} i_{p_m} - v_{p_m} \tag{15}$$

$$ARR_6 = \left(V_{oc} - v_{C_{dl_m}}\right) - v_{b_m} - R_{i_n} i_{b_m} \tag{16}$$

$$a_1 = \left|\frac{3}{2}\delta_{L_q} pL_{q_n} i_{q_m}\Omega_m\right| + \left|\delta_{L_d} L_{d_n}\frac{di_{d_m}}{dt}\right| + \left|\delta_{R_d} R_{d_n} i_{d_m}\right| \tag{17}$$

$$a_2 = \left|\frac{3}{2}\delta_{L_d} pL_{d_n} i_{d_m}\Omega_m\right| + \left|\delta_{L_q} L_{q_n}\frac{di_{q_m}}{dt}\right| + \left|\delta_{R_q} R_{q_n} i_{q_m}\right| \tag{18}$$

$$a_3 = \left|\delta_J J_n\frac{d\Omega_m}{dt}\right| + \left|\delta_\sigma\sigma_n\Omega_m\right| \tag{19}$$

$$a_4 = \left|\delta_{R_{eq}} R_{eq_n}\left(i_{ph} - i_{p_m}\right)\right| + \left|\delta_{R_s} R_{s_n} i_{p_m}\right| \tag{20}$$

$$a_6 = \left|\delta_{R_i} R_{i_n} i_{b_m}\right| \tag{21}$$

After deriving the *ARRs* from the LFT-BG model of HMS, the Fault signature matrix (FSM) illustrated in table 1 and table 2 are performed on the obtained *ARRs*. they consist of the fault signatures used to carry out the monitorability M_b and isolability I_b analysis as follows: $M_b=1$ if the fault is sensitive at least to one *ARR*. $M_b=0$ in the opposite case. $I_b=1$, if the fault has a unique signature. Else, $I_b=0$ [6].

Table 1 and table 2 show that any fault cannot be isolated separately. However, the presented analysis shows that the faults can be isolated by sets depicted by specific colors. Each set represents the same signature. This way can, at least, localize the fault at the specific HMS part where they occur.

Table 1. Fault signature matrix, M_b: monitorability index, I_b: isolability index.

Faults→ ARRS↓	R_d	v_{dm}	L_q	L_d	i_{dm}	i_{qm}	Ω_m	R_d	v_{pm}	J	σ	R_s	R_{sh}	v_{pm}	i_{pm}	C_{dl}	R_{dl}	i_{bm}
ARR_1	1	1	1	1	1	1	1	0	0	0	0	0	0	0	0	0	0	0
ARR_2	0	0	1	1	1	1	1	1	1	1	1	0	0	0	0	0	0	0
ARR_3	0	0	0	0	0	1	1	0	0	0	0	0	0	0	0	0	0	0
ARR_4	0	0	0	0	0	0	0	0	0	0	0	1	1	1	1	0	0	0
ARR_5	0	0	0	0	0	0	0	0	0	0	0	0	0	0	0	1	1	1
ARR_6	0	0	0	0	0	0	0	0	0	0	0	0	0	0	0	0	0	1
M_b	1	1	1	1	1	1	1	1	1	1	1	1	1	1	1	1	1	1
I_b	0	0	0	0	0	0	0	0	0	0	0	0	0	0	0	0	0	0

Table 2. Fault signature matrix carried out on the faults may occur in R_i, v_{bm}, v_{cdlm}.

Faults→ ARRS↓	R_i	v_{bm}	v_{cdlm}
ARR_1	0	0	0
ARR_2	0	0	0
ARR_3	0	0	0
ARR_4	0	0	0
ARR_5	0	0	1
ARR_6	1	1	1
M_b	1	1	1
I_b	0	0	0

4. Simulation study

The simulation study is carried out to validate the robust monitoring for FDI of HMS in the presence of system uncertainties. The simulation consists of implementing the *ARRs* with their adaptive thresholds and the *HMS* model under the operation state in the 20sim software, where the parameter random variations (HMS parameters variate in the interval of values related to the relative uncertainties) are included in the HMS model to simulate the parameter uncertainties. One scenario is performed to validate the proposed approach, which consists of injecting the faults in certain parameters: R_d, Ω_m (sensor fault), σ, R_s, and R_i at different times: 10s, 30s, 38s, 45s, and 50s, respectively. There is in the simulation a process used to cancel the injected faults to let the faults affecting the system during a period of time. The simulation results are presented in figure 10 and figure 11.

According to the obtained results, the residuals signals generated from the developed *ARRs* are not in the zero value and vary randomly due to the parameter uncertainties. Their thresholds are adaptively evolving with the evolution of system variables. The adaptive evolution of the thresholds helps to prevent the residual signals from going out of their thresholds bounds when there is no fault. When faults occur in the HMS, the residuals generated by ARR_1, ARR_2, ARR_3, ARR_4, and ARR_6 go out of their thresholds bounds at different times due to the presence of faults. When the faults are canceled, the residual signals generated by the *ARRs*, which are previously mentioned, return to the threshold bounds. It is noticed that only the residual signal generated by ARR_5 is not affected and kept inside its thresholds bound during the whole simulation time. According to the monitorability and isolability carried out in section 3, ARR_5 is not sensitive to any fault simulated in this scenario. For this reason, the residual signal of ARR_5 is not affected by these faults. This part of the validation scenario and the presence of faults at different times have been performed to verify the monitorability and isolability analysis discussed in section 3 based on the FMS technique.

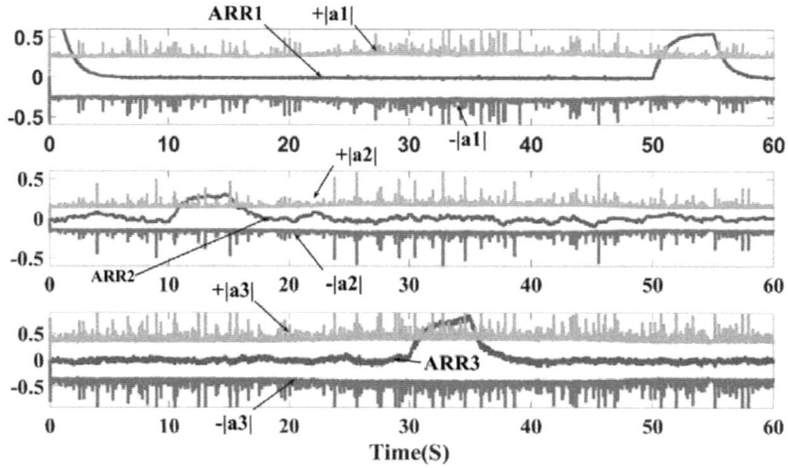

Figure 10. Numerical evaluation of the residuals: *r1*, *r2*, *r3* with their adaptive thresholds.

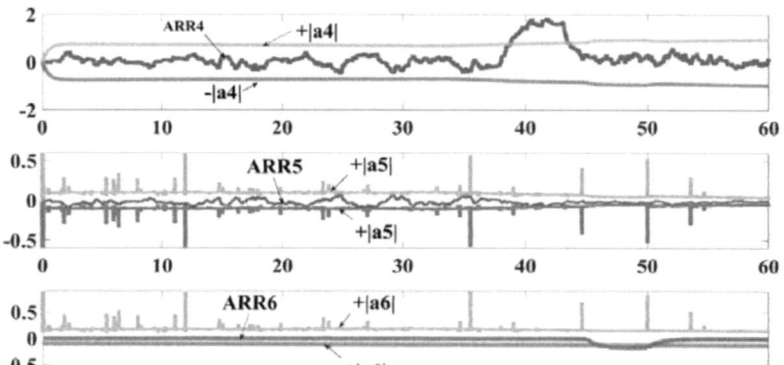

Figure 11. Numerical evaluation of the residuals: *r4*, *r5*, *r6* with their adaptive thresholds.

5.Conclusion

The obtained results prove the LTF-BG modeling approach's effectiveness in designing the online robust monitoring to detect and isolate faults in the HMS. The online FDI process earns the robustness from the adaptive thresholds, which evolve with the evolution of the system variables in the presence of parameter uncertainties. Furthermore, the indicators of faults under this approach are only sensitive to faults. This study shows the LFT-BG approach's ability to model with accuracy way the HMS. As LFT-BG allows all phenomena acting negatively on the system to take part in the model, dealing with these issues, as mentioned earlier, becomes easy when LFT-BG is used, and improving the fault isolation capability under single and multiple faults requires more sensors of monitoring that will be the future work.

6. References

[1] Abdallah I, Gehin A L and Ould Bouamama B 2018 Event driven hybrid bond graph for hybrid renewable energy systems part I: Modeling and operating mode management International journal of hydrogen 43 22088-22107
[2] Parakash O, Samantaray A K and Bhattacharyya 2018 Model-based multi-component adaptive prognosis of hybrid dynamical systems Control engineering practice 72 1-18
[3] zhong M, Song Y and Ding S X 2015 Parity space-based fault detection for linear discrete time-varying systems with unknown inputs Automatica 59 120-126
[4] Alhelou H H, Golshan M E H, Hatziargyriou N D and Moghaddam M P 2020 IEEE Transactions on Industrial Informatics 1 - 1
[5] Abdallah I, Gehin A L and Ould Bouamama B 2018 On-line robust graphical diagnoser for hybrid dynamical systems Engineering applications of artificial intelligence 69 36-49
[6] Samantaray A K and Ould Bouamama B 2008 Model-based Process Supervision Springer *book*
[7] Jha M S, Tanguy G D and Ould Bouamama B 2018 Robust fault detection with Interval Valued Uncertainties in Bond Graph Framework Control Engineering Practice 71 61-78
[8] Rahal M I, Ould bouamama B and Meghebbar A 2013 Hybrid Bond Graph for robust diagnosis - Application to hydraulic system 3rd International Conference on Systems and Control.
[9] Touati Y, Merzouki R and Ould Bouamama B 2012 Robust diagnosis to measurement uncertainties using bond graph approach: Application to intelligent autonomous vehicle Mechatronics 22 1148-1160
[10] Liao M, Dong L, Jin L and Wang S 2009 Study on Rotational Speed Feedback Torque Control for Wind Turbine Generator System International Conference on Energy and Environment Technology
[11] Yurika, Suwarno, Sianipar G H and Naiborhu J 2019 Modelling of Synchronous Generator for Transient Stability in Power System 2019 2nd International Conference on High Voltage Engineering and Power Systems
[12] Alias K, Azah M, Hussain S, Afida A and Aras M S M 2013 Modelling and Simulation of a Single Phase Grid Connected Using Photovoltaic and Battery Based Power Generation European Modelling Symposium

A Triple-threshold-based Load Event Detection Algorithm for Non-intrusive Load Monitoring

Mingming Chen[1,2], Kaijie Fang[1,2], Qifeng Huang[1,2], Shihai Yang[1,2], Hanmiao Cheng[1,2,*], Tianchang Liu[1,2], Yixuan Huang[1,2]

[1]State Grid Jiangsu Electric Power Co., LTD Marketing Service Center, China
[2]State Grid Laboratory of Power Demand Side Management Technology, Nanjing, China

chenghm1@js.sgcc.com.cn

Abstract. Event detection is an important foundation of non-intrusive load monitoring algorithm. In this paper, the common household appliance load events are classified, and a new triple-threshold event detection algorithm is proposed aimed at solving the problems of false detection and missing detection in the practical application. Firstly, a low power threshold is used to realize high-sensitive detection of the load events, and secondly the detected events are spliced according to the time threshold to get the complete events. Thirdly, the high threshold is used to discriminate the complete event set to filter out the disturbance caused by load fluctuation. Finally, the results are modified with a correction logic. The test results carried with static data show that, the algorithm proposed in this paper is more accurate for positioning the time of putting into and cutting off load, which is conducive to improve the accuracy of transient interval interception of load events, and has advantages in detecting slow rising load events. In addition, the algorithm proposed in this paper has a small amount of calculation, which can meet the requirements of application in the hardware of smart meter.

1. Introduction

Non-intrusive load monitoring (NILM) is a new technique to obtain residential power consumption details. This technique uses smart meter sampling data, applies signal analysis and machine learning algorithm, so that it can obtain some information such as start-up and stop time, energy consumption and usage rule of electrical appliances[1,2]. The implementation steps of NILM generally include data acquisition, data processing, event detection, feature extraction, and load classification[3]. The specific details of each step are as follows: a) data acquisition is used for digital sampling of analog voltage and current signal; b) data processing is used to calculate root mean square (RMS), power, harmonic and other electrical parameters; c) event detection is used to detect electrical switch or state switching events from data sequence; d) feature extraction and load classification have a strong correlation and are used to assign categories to the detected electrical appliances. In the above steps, event detection is the most important. Almost all NILM algorithms need to distinguish the steady-state and transient sections before and after load switching by event detection. On this basis, the subsequent calculation is performed. Event detection algorithm has a great influence on the performance of NILM algorithm, even determines the success or failure of the design of NILM algorithm.

There are many research results about load event detection algorithm. In the Ref. [4], a sliding window based cumulative sum (CUSUM) algorithm is proposed, which can effectively detect load switching events. In the Ref. [5], the research results of load event detection algorithm , including multi decision fusion [6], pattern recognition [7], improved Canny algorithm [8], wavelet transform [9], etc, are fully summarized. On this basis, a method combining adaptive Gaussian filter with CUSUM is designed, which is called improved CUSUM algorithm. By introducing Gaussian filter, the value of β in the Ref. [4] is reduced, so as to improve the detection accuracy. In the Ref. [10], a sliding window event detection algorithm based on power variance judgment, called two-step variance method, is proposed. This algorithm can determine the switching time of electrical appliances and avoid the interference of system noise and small power fluctuation. The above algorithms have follow problems in practical application: a) The detection rate of bilateral CUSUM for slow varying load events is low, and its anti-noise performance is insufficient; b) The improved CUSUM improves the noise immunity, but it has no obvious effect on the detection of slow varying load. Moreover, if the value of h in the Ref. [5] is increased, the transient process interception time error will be increased; c) The two-step variance method is not suitable for complex appliance load mode transformation, and it can not detect slow varying load events. In practical application, there are many slow varying loads, and the typical load is variable frequency air condition. The existing event detection algorithms have the problem of missing detection when dealing with variable frequency air condition.

In order to solve some problems of event detection algorithm in practical application, this paper proposes a new idea of load event detection. Firstly, low threshold is used to detect load events with high sensitivity, and then load events are spliced according to time threshold. Finally, high threshold is used to screen effective load events, which can ensure the detection performance of conventional load events and detect slow varying load events accurately. Simulation results show that, compared with the existing algorithms, the proposed method improves the detection accuracy of conventional load events, and significantly improves the detection accuracy of slow varying and impact events.

2. Classification of residential load events

Residential power load event detection is essentially the problem of data sequence change point detection. In this paper, the power sequence is used as the object of load event detection. With the development of economy and technology, there are many kinds of household appliances, and their working characteristics and start stop characteristics are very different. In this paper, the different load events are classified firstly.

2.1 Step-type load event

This kind of event is mainly caused by the switching of pure resistive electrical appliances in the family, and the power curve is similar to the step function, such as electric kettle, electric water heater, electric rice cooker, electric radiator, etc. The Fig.1 shows the measured power curve when the electric kettle is turned on.

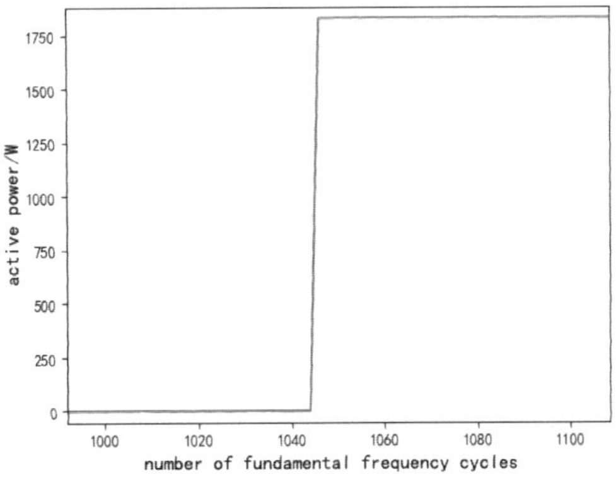

Figure.1 Power curve of the electric kettle

It can be seen from the Fig.1 that the step-type load event has two main characteristics: a) the power fluctuation before and after load event is small, generally less than 10W; b) the transient time of switching on and off is very short, which generally does not exceed 60ms.

2.2 *Impact load event*

This kind of event is mainly caused by electrical appliances with compressor or motor, such as air conditioner, vacuum cleaner, refrigerator, etc. The Fig.2 shows the measured power curve when the vacuum cleaner is turned on.

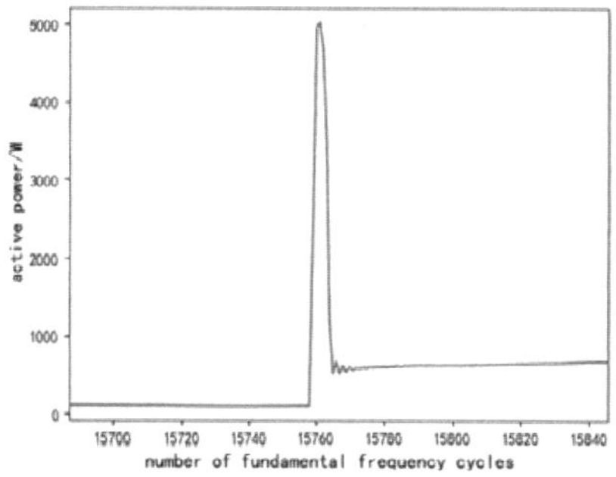

Figure.2 Power curve of the vacuum cleaner

It can be seen from the Fig.2 that the impact load event has two main characteristics: a) there is a large impulse after the load is put into operation, and the peak power is usually several times or even tens of times of the steady-state power; b) the impact time is short, and the time width of the impact peak is about 0.2S to 0.6s.

2.3 *Slow rising load event*

This kind of event is mainly caused by electrical appliances with inverter components, such as induction cooker and variable frequency air conditioner. The starting is a long-lasting power rising process, and the duration is related to the control strategy. The measured starting power curves of induction cooker and variable frequency air conditioner are respectively shown in the Fig.3 and Fig.4.

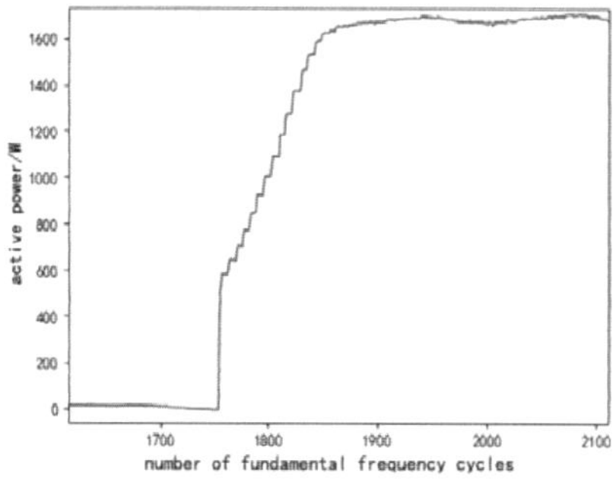

Figure.3 Power curve of the electromagnetic furnace

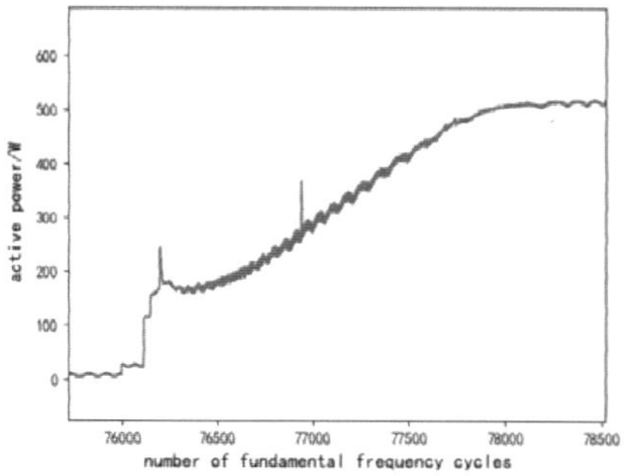

Figure.4 Power curve of the inverter air conditioner

It can be seen from the Fig.3 and the Fig.4 that the duration of slow rising load events is generally several seconds or even minutes, and it is accompanied by a small amplitude of power fluctuation in the process of power rising, showing a saw-tooth shape.

3. Principle of triple-threshold load event detection algorithm

In view of the shortcomings of the existing algorithms for load event detection, the basic idea of the triple-threshold algorithm proposed in this paper is to divide event detection into three steps: a) detect mutation events using low threshold, and consider the small load fluctuation as a part of load events; b) splice the abrupt events according to the time threshold, that is, the load events in a short time are considered to be a transient process; c) pick up the effective load events using high threshold, so as to filter out the disturbance caused by load fluctuation. Taking the event of putting load into operation as an example, the algorithm implementation steps are described in detail.

3.1 S1: detecting mutation point using low threshold

According to the sampling data of current and voltage, the power value is calculated and the power sequences are obtained. The sliding window sequence located at power point P_i is notated as

$$SW_i = \{SW_i^{\text{pre}}, P_i, SW_i^{\text{post}}\} \tag{1}$$

Where, SW_i^{pre} is defined as forward window, and SW_i^{post} is defined as backward window, which represent the power sequence with width w before and after P_i respectively. SW_i^{pre} and SW_i^{post} are defined as

$$\begin{cases} SW_i^{\text{pre}} = \{P_{i-w}, P_{i-w+1}, \cdots, P_{i-1}\} \\ SW_i^{\text{post}} = \{P_{i+1}, P_{i+2}, \cdots, P_{i+w}\} \end{cases} \tag{2}$$

The low threshold LOW_{thr} is used to detect whether P_i is a mutation point. The discriminant is

$$\Delta P = \mu(SW_i^{post}) - \mu(SW_i^{pre}) \tag{3}$$

Where, $\mu(\cdot)$ represents the mean function. If $\Delta P > LOW_{thr}$, then P_i is considered as a mutation point. Assuming that the power point P_i is a mutation point, the close set composed of the index of the power sequence number of the sliding window is

$$D_{up}(i) = [i-w, i+w] \tag{4}$$

In the first step, setting a low threshold can help to improve the sensitivity of event detection and minimize missed detection, but it will inevitably lead to the problem of detecting additional events , which need to be solved by event splicing.

3.2 S2: splicing events using time threshold

For a slow rising load event, the transient process will last for a long time. Because of the low threshold, more than one event are inevitably detected out. Therefore, it is necessary to splice these events into one event according to the time threshold. As shown in the Fig.5, in the whole start-up process of a variable frequency air conditioner, several mutation points will be detected out in multiple sliding windows. That is to say, more than one event will be detected out in the start-up process. In the second step, these additional events will be spliced into one event.

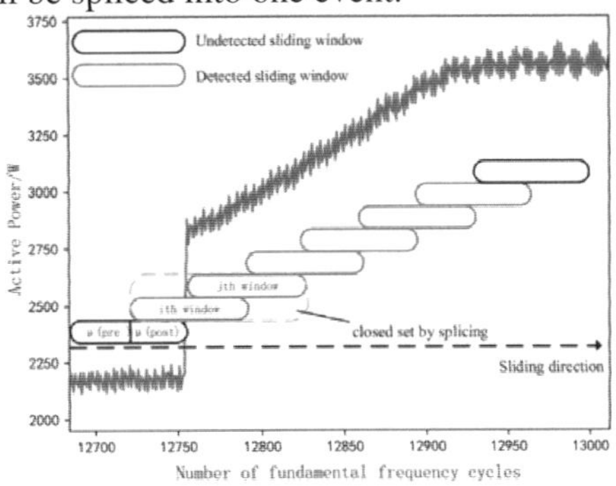

Figure.5 Power curve of the inverter air conditioner

In the Fig.5, the ith and jth sliding windows are taken as examples to illustrate the load event splicing process. The set formed by the index of the ith sliding window is denoted as $D_{up}(i)$. If

$D_{up}(i) \cap D_{up}(j) \neq \varphi$, then it is considered that the ith and jth sliding windows are in the transient process of the same load event. The discriminant is as follows

$$\Delta T = j - i - 2w \tag{5}$$

The time threshold is set to $TIME_{thr}$. If $\Delta T \leq TIME_{thr}$, then the load mutation points detected in the two sliding windows are considered to belong to the transient process of the same event. Merge $D_{up}(i)$ and $D_{up}(j)$ into a closed set:

$$D_{event}(i) = [L(i), R(i)] \tag{6}$$

Where, $D_{event}(i)$ is the ith load event, $L(i)$ and $R(i)$ are

$$\begin{cases} L(i) = i - w \\ R(i) = j + w \end{cases} \tag{7}$$

After the step of event splicing, all power mutation points detected in the same long transient process will be spliced into the same event. Finally, the sequence of events obtained is $\{D_{event}(1), D_{event}(2), \cdots, D_{event}(N)\}$.

3.3 S3: filtering load events using high threshold

High sensitivity detection and time-domain splicing, which can detect slow rising events, but can not overcome the problem of false detection causing by load fluctuation. Therefore, a high threshold $HIGH_{thr}$ shall be used to filter the event sequence obtained in the second step. To determine whether $D_{event}(i)$ is an event, the discriminant is defined as follow

$$\Delta\overline{P} = \frac{1}{w}\sum_{k=L(i)}^{L(i)+wl} P_k - \frac{1}{w}\sum_{k=R(i)-wl}^{R(i)} P_k \tag{8}$$

Where, if $\Delta\overline{P} > HIGH_{thr}$, then $D_{event}(i)$ is considered to be a real load event rather than a fluctuation disturbance.

4. Experimental test and analysis

In order to verify the effectiveness of the detection algorithm proposed in this paper, the improved CUSUM algorithm and two-step variance detection method are selected as the comparison objects, and the three algorithms are tested and compared by using the actual electrical load static data. This test method has the convenience of simulation, and can reflect the actual performance. The relevant parameters are set as follows: $LOW_{thr} = 5\text{W}$, $TIME_{thr} = 0.8\text{s}$, $HIGH_{thr} = 20\text{W}$, $w = 0.4\text{s}$, the sliding window distance is the same as w, other parameters are consistent with those in the comparative references.

4.1 Example 1: step-type load event detection

In this example, the electric kettle is used as the tested electrical appliance of which the nominal power is 1800W. The electric kettle is controlled to start and stop for many times. The detection results of step load events with three detection algorithms are shown in the Fig.6, Fig.7 and Fig.8. The red dot indicates the active power curve segment of load events detected by the algorithm.

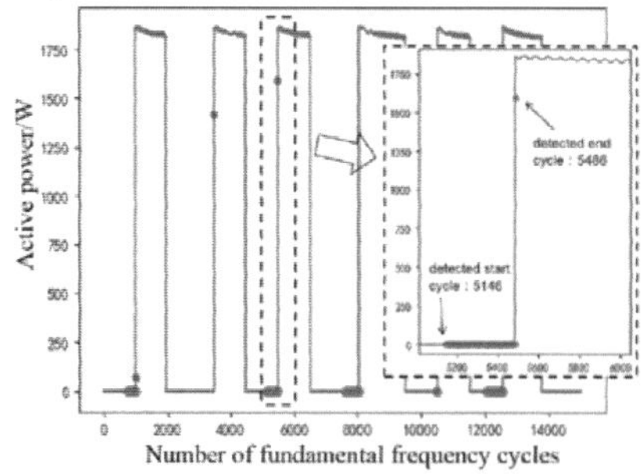

Figure.6 Detection results of two-step variance method

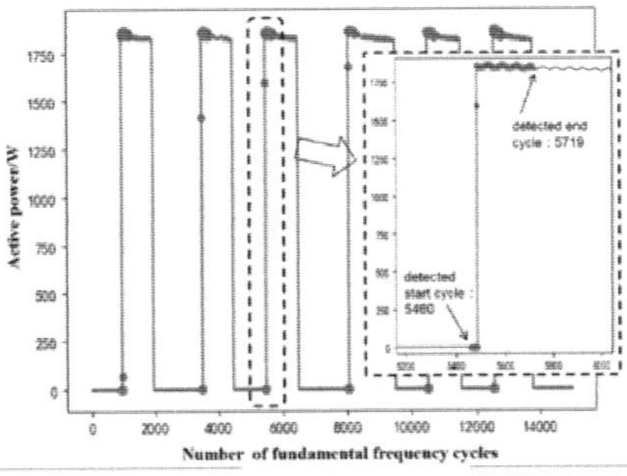

Figure.7 Detection results of improved CUSUM algorithm

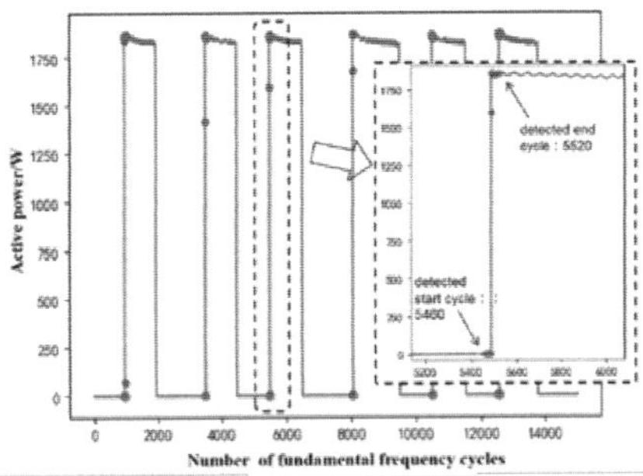

Figure.8 Detection results of triple-threshold algorithm

Comparing with the Fig.6, Fig.7 and Fig.8, it can be seen that: a) The two-step variance method can detect the beginning position of the event in advance, but loss a segment at the end of the event; b) The improved CUSUM algorithm can detect the start position of the event accurately, but the delay of the end position of the event is large; c) The algorithm proposed in this paper can accurately detect the starting position of the event.

4.2 Example 2: impact load event detection

In this example, the fixed frequency air conditioner is used as the tested electrical appliance. The results of the three detection algorithms are shown in the Fig.9, Fig.10 and Fig.11. The red dot indicates the active power curve segment of load event detected by the algorithm.

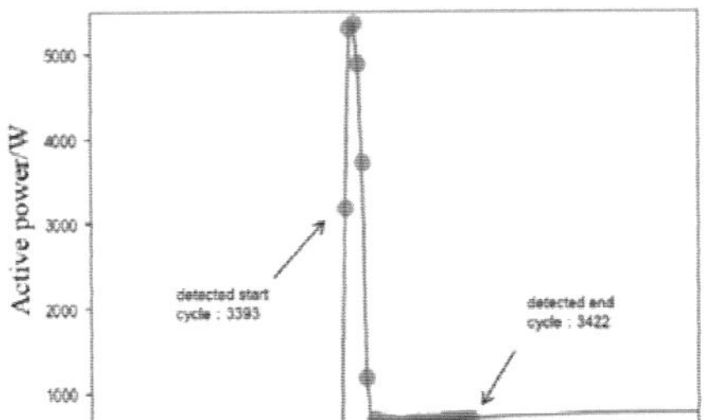

Figure.9 Detection results of two-step variance method

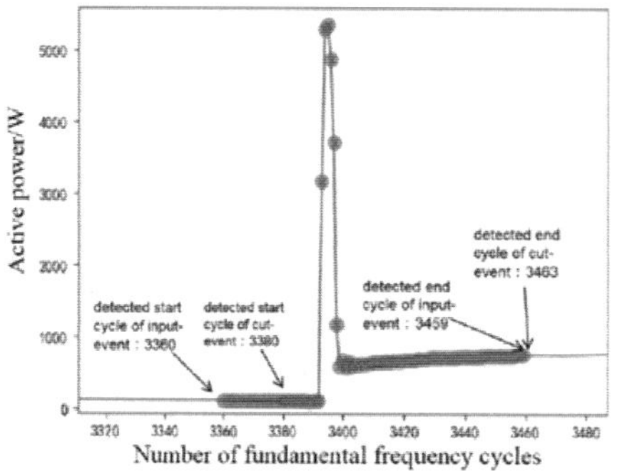

Figure.10 Detection results of improved CUSUM algorithm

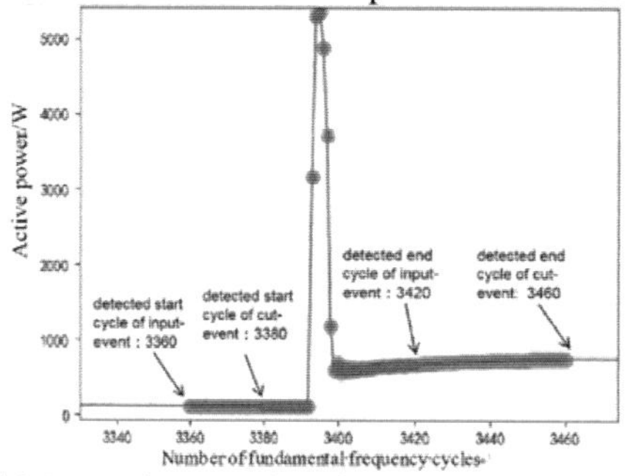

Figure.11 Detection results of triple-threshold algorithm

Comparing with the Fig.9, Fig.10 and Fig.11, it can be seen that: a) Two-step variance method has a good detection effect, but it still has the problem of large position error; b) A turning-off event will be detected in the descending section of the peak since the improved CUSUM algorithm is a bilateral cumulative calculation.

4.3 *Example 3: slow rising event detection*

Variable frequency air conditioner is a typical slow rising load, and its slow rising time varies from a few seconds to a few minutes, which is a great challenge to the detection algorithm. In this example, the inverter air conditioner is used as the tested electrical appliance. The input power of refrigeration is 1068w, the heating power is 1120w, and the maximum current is 12.2A.

The first two detection algorithms can not detect the load event during the slow rising process of the air conditioner. The detection results of the proposed algorithm are shown in the Fig.12.

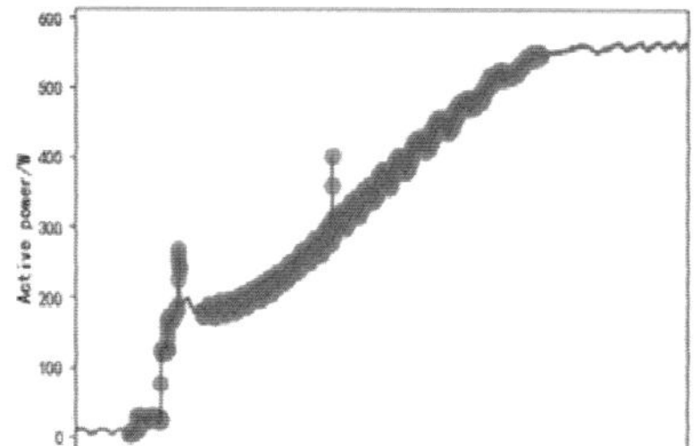

Figure. 12 Detection results of slowly rising load events

As it can be seen from the Fig.12, the duration of the tested variable frequency air conditioner from start to steady state is about 40 seconds. Due to the slow rising speed of power, the variance value and cumulative value in the short-term sliding window are very small, so the first two algorithms can not detect the start of variable frequency air conditioner, which will directly lead to the failure of NILM algorithm based on event detection to identify the variable frequency air conditioner. The algorithm proposed in this paper first detects a series of continuous events in the slow rise section with low threshold, and then splices the events with time threshold. The power change of the slow rising event formed by splicing is about 600W, which obviously can pass the high threshold screening, and finally detects turning-on event of the frequency conversion air conditioning.

5. Conclusions

NILM algorithm is highly dependent on event detection. In this paper, a new idea of event detection algorithm is proposed based on engineering practice. The algorithm has good performance in practice through combining detecting, splicing and filtering according different thresholds.

In this paper, the detection performance is tested with the actual static data. The results show that the algorithm has more advantages than the existing algorithms in the detection of slow rising load events, and has been applied in a large-scale pilot project in Jiangsu Province.

The algorithm proposed in this paper also inevitably has false detection and missed detection, and there is still space for optimization of multi threshold cooperation. Taking the detection accuracy, recall rate and other parameters as the optimization objectives, multi threshold combination optimization can be carried out, which has not been discussed in this paper.

Acknowledgment

This work was supported by State Grid Corporation of State Grid Jiangsu Electric Power Co., Ltd. Project "Research on non-intrusive residential electricity information perception technology and module development (JF2021010)". The authors also would like to thank the reviewers and editors for their generous help.

6. References

[1] G. W. Hart, 1992, "Non-instructive appliance load monitoring", Proceedings of the IEEE, vol.80, no.12, 1870-1891.

[2] S. Singh and A. Majumdar, 2020, "Non-Intrusive Load Monitoring via Multi-Label Sparse Representation-Based Classification", IEEE Transactions on Smart Grid, vol.11, no.2, 1799-1801.

[3] K. D. Anderson, 2014, "Non-Intrusive Load Monitoring: Disaggregation of Energy by Unsupervised Power Consumption Clustering", USA, Carnegie Mellon University.

[4] Niu Lulu, Jia Hongjie, "Transient Event Detection Algorithm for Non-Intrusive Load Monitoring", Automation of Electric Power System, 2011, vol.35, no.9, 30-35.

[5] Qu Hedi, Li Jian, Zhang Zhihao, et al., 2018, "Study of Fast Identification Edge Detection Method for Electrical Equipment", Proceedings of the CSEE, vol.38, no.15, 4566-4575.

[6] Wang Xiaohuan, Li Ruyi, Zhou Dongguo, et al., 2016, "Non-intrusive power load disaggregation method based on decision fusion and its applications", Power System Protection and Control, vol.44, no.7, 115-121

[7] B. Liu, 2014, "Non-Intrusive Load Monitoring and Disaggregation Technology", China, Tianjin University.

[8] Abdullah-Al-Nahid, Kong Yinan, Hasan M N, 2015, "Performance analysis of Canny's edge detection method for modified threshold algorithms", Proceedings of 2015 International Conference on Electrical & Electronic Engineering(ICEEE), Rajshahi, Bangladesh, IEEE, 93-96.

[9] Li Ruyi, Huang Mingshan, Zhou Dongguo, et al., 2016, "Optimized nonintrusive load disaggregation method using particle swarm optimization algorithm", Power System Protection and Control, vol.44, no.8, 30-36.

[10] Zhou Ming, Song Xufan, Tu Jing, et al., 2018, "Residential Electricity Consumption Behavior Analysis Based on Non-Intrusive Load Monitoring", Power System Technology, vol.42, no.10, 3268-3274.

Study on the Electrical Conductivity of Inorganic Conductive Network in Sediments

R Y Zhao[1], Z W Zhao[1,*], Z J Weng[1], Y Fang[1] and H L Jiang[2]

[1] School of Electronic Science and Engineering, Southeast University, Nanjing 210096, P. R. China
[2] State Key Laboratory of Lake Science and Environment, Nanjing Institute of Geography and Limnology, Chinese Academy of Sciences, 73 East Beijing Road, Nanjing 210008, P. R. China

E-mail: Zhao_zw@seu.edu.cn

Abstract. The inorganic conductive network provides an essential channel for electron transport and supports the biogeochemical process in sediments, but the conductive mechanism of conductive network is not well understood. In this work, theory of circuit and electronics was applied to build a three-dimensional (3D) resistivity network simulation model for exploring the conductive mechanism and analysing the effect of the particle size on the conductive characteristic of inorganic conductive network. In order to simulate the real sediment environment, inorganic composites with silica (SiO_2) particles as matrix using magnetite (Fe_3O_4) particles as fillers are constructed. The simulation results reveal that the electrical conductivity of these composites rises nonlinearly with the increasing volume fraction of conductive fillers, which is consistent with the percolation theory. Moreover, small-sized conductive particles or large-sized matrix particles are confirmed to exert a positive part in enhancing electrical conductivity of composites.

1. Introduction

It was found that the spatially separated redox process coupled by bioelectric current exists in sediments through the conductive networks composed of complex inorganic and organic components [1]. Compared with organic components, mineral particles, abundant in sediments and considered as the main inorganic conductive components that constitute conductive network in sediments, such as Fe_3O_4 and FeS_2, not only participate in electron transmission between microbial species, but also significantly enhance the conductivity of sediments [2-4].

In addition, the mechanism of long-distance electron transmission in sediments provides a novel solution and scientific reference for systematically regulating the biogeochemical process of water elements, sediment remediation and developing new energy materials [5-6]. However, in previous work, studies on long-distance electron transport in sediments mainly focus on biological and environmental fields, several researchers have attempted to investigate the construction mechanism and electrical conductivity of conductive network in sediments.

The percolation theory and quantum tunnelling theory [7-8], widely used to investigate polymer composites [9-10], is applied to the study of inorganic conductive networks in sediments in this work. Based on these theories, resistance composition and distribution law of the inorganic conductive network simulated by composite material with inorganic SiO_2 particles as matrix and Fe_3O_4 particles as conductive fillers are analysed. After that, a 3D simulation model is established to calculate the electrical conductivity of conductive composites based on the Kirchhoff's Current Law (KCL) and investigate the effects of particle size on the electrical conductivity of Fe_3O_4/SiO_2 composites.

2. Simulation of conductive network

2.1. Establishment of simulation model

Spheres are used to simulate the SiO_2 particles and Fe_3O_4 particles, and the hard-core model [11-12] is chosen to build a single particle model in that it is more exact when considering tunnelling effect [13]. As shown in figure 1(a), the particle consists of an impenetrable solid core with radius of r and a penetrable soft shell with thickness of $t/2$. The solid core represents the particle entity, while the outer shell represents the effective tunnelling effect area. If the distance d between two adjacent particles is less than the effective tunnelling effect distance t [14], particles will be regarded as connected with each other by tunnelling conduction. Otherwise, the path among two particles is an open circuit.

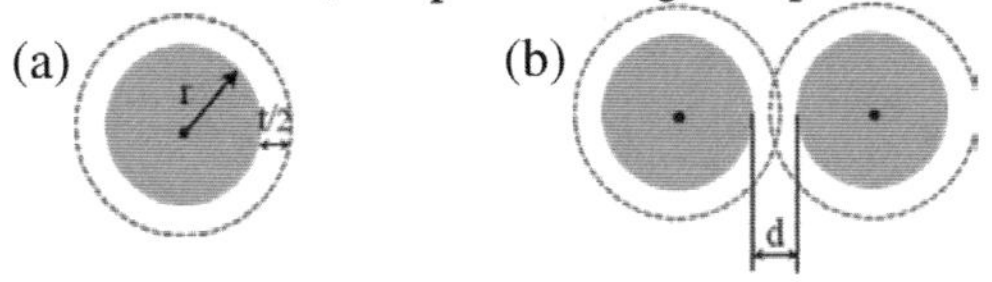

Figure 1. (a) Single particle hard-core model. (b) Adjacent particles model

The length, width and height of the simulation generated domain is defined as L, W and H respectively. The spatial location of particle randomly distributing in 3D space is expressed by the equation:

$$\begin{cases} x = (L \cdot rand) - (\frac{L}{2}) \\ y = (W \cdot rand) - (\frac{W}{2}) \\ z = (H \cdot rand) - (\frac{H}{2}) \end{cases} \tag{1}$$

Where *rand* is a random number between [0, 1] region, and (x, y, z) is the centre point coordinate of particle. When the distance between two particles is smaller than t, the electrons between two separated particles can pass through the potential barrier to realize electron transmission [12]. In this case, the resistance between two particles consists of the intrinsic bulk resistance of conductive particles and the tunnelling junction resistance [15] between particles caused by electron tunnelling. Beyond this distance, the value of resistance approaches infinity. Therefore, the resistance R_{ij} between the *i-th* particle and *j-th* particle depends on the distance d_{ij} between two particles is written by the equation:

$$R_{ij} = \begin{cases} R_i + R_t\,, & 0 < d_{ij} < t \\ \infty\,, & d_{ij} \geq t \end{cases} \tag{2}$$

Where R_i is the intrinsic bulk resistance of particle, R_t is the tunnelling junction resistance between the *i-th* particle and the *j-th* particle. The tunnelling junction resistance is calculated by Simmons formula [16-17], which is expressed by the following equation:

$$R_t = \frac{U_{ij}}{AJ} = \frac{P^2 d_{ij}}{Ae^2\sqrt{2m\lambda}} \exp(\frac{4\pi d_{ij}}{P}\sqrt{2m\lambda}) \tag{3}$$

Where J is the tunnelling current density, U_{ij} is the electrical potential difference of particles, A is the cross-sectional area which is assumed to be the same as the cross-sectional area of particles [12,17], e is the single electron charge, m is the electron mass, P is Planck constant, and λ is the barrier height. The intrinsic bulk resistance of particles calculated by the volume method can be obtained:

$$R = \frac{1}{\int \frac{1}{dR}} = \frac{1}{\int_0^r \frac{\pi R dR}{\rho\sqrt{r^2 - R^2}}} = \frac{\rho}{\pi r} \tag{4}$$

Where ρ is the resistivity of the particle and r is the particle radius.

2.2. Solution of simulation model

Firstly, the centre of each particle is defined as a node, and the node voltage analysis method [18] is used to solve the voltage of any node on the circuit. Selecting one node as a zero-potential reference point, and voltage u_i at the *i-th* node is equal to the potential difference between the *i-th* node and the reference point. For the *i-th* node, the current flowing into the node can be given as follows:

$$I_i = u_i \sum_{k=1}^{n} g_{ki} - \sum_{k=1}^{n} u_k g_{ki} \quad , \ k \neq i \tag{5}$$

Where g_{ki} is the conductance between the *k-th* node and the *i-th* node. It is assumed that the lower boundary is zero-potential reference point and the upper boundary is observation point, and the node numbers of these boundaries are set to *n* and *1*, respectively. The current flowing into the upper boundary is set to 1A, which is equal to the corresponding current flowing out of the lower boundary. Since the other nodes inside the conductive network are not in contact with the outside, the currents flowing through these nodes inside are zero. According to the above analysis, the node current equation set can be obtained:

$$\begin{bmatrix} \sum g_{1,\dots} & -g_{12} & \cdots & -g_{1(n-1)} \\ -g_{21} & \sum g_{2,\dots} & \cdots & -g_{2(n-1)} \\ \vdots & \vdots & \ddots & \vdots \\ -g_{n-1,1} & -g_{n-1,2} & \cdots & \sum g_{n-1,\dots} \end{bmatrix} \begin{bmatrix} u_1 \\ u_2 \\ \vdots \\ u_{n-1} \end{bmatrix} = \begin{bmatrix} 1 \\ 0 \\ \vdots \\ 0 \end{bmatrix} \tag{6}$$

Where $g_{ij} = g_{ji}$, and $\sum g_{i,\dots}$ is the sum of all the conductance connected with the *i-th* node. The minimum norm solution method (using minres function in Matlab software) is applied to solve the linear equations. Subsequently, the equivalent resistance and electrical conductivity of the conductive network can be solved according to Ohm's law.

3. Results and discussion

3.1. Conductive mechanisms of composites

It has been proved that a classical power law formula [19], i.e., $\sigma = \sigma_0(\phi - \phi_c)^t$, can exactly reflect the electrical percolation behaviour of the particle composites. Here exponent t is a constant, ϕ is the volume fraction of conductive particles and ϕ_c is the percolation threshold which is equal to the critical value of volume fraction of conductive particles at the sharp rise of electrical conductivity. The electrical conductivity of the composites (i.e., σ) is directly proportional to the conductivity of conductive particles (i.e., σ_0). In table 1, ϕ_c is calculated by fitting the power law formula to the simulation results in figure 2 for the composites with different particle size of Fe_3O_4 in SiO_2 matrix. Herein, similar percolation behaviour is observed in figure 2 for all composites and the electrical conductivity curve shows a steep rise near the expected percolation threshold, which show that the proposed model can predict the percolation behaviour of composites correctly.

Table 1. Percolation threshold obtained by fitting the power law formula

Composites	Percolation threshold ϕ_c (vol%)
Fe_3O_4 (20nm)/SiO_2	17.01
Fe_3O_4 (200nm)/SiO_2	22.13
Fe_3O_4 (2μm)/SiO_2	25.64

As shown in figure 2, the volume fraction of conductive fillers significantly affects the electrical properties of composites [12, 17-18]. Taking Fe_3O_4(20nm)/SiO_2 composite as an example, the electrical conductivity curve increases rapidly near the percolation threshold $\phi_c = 17.01\%$, which is consistent with the percolation formula above. Initially, the electrical conductivity is not influenced by a small quantity of particles. At this time, the average distance between conductive particles is larger than the effective distance of tunnelling effect, and thus the path among the adjacent particles is an open circuit, and the contact resistance between these particles approaches to infinite. Therefore, the composites mainly represent the weak conductivity of the matrix. When the volume fraction of fillers ranges from 17.01% to 30%, the trend of electrical conductivity rises dramatically from 8×10^{-14} S/m to

Journal of Physics: Conference Series **2065** (2021) 012012 doi:10.1088/1742-6596/2065/1/012012

5×10^{-2} S/m. During this process, the electron tunnelling between particles is easier to realize owing to the ever-decreasing average distance of fillers. With the help of electron tunnelling between conductive fillers, conductive paths will be formed between adjacent conductive particles, and the series-parallel action of conductive paths will form a network for transmitting electrons, which makes the composites show electrical conductivity. Once the volume fraction of fillers arrives at 30%, the electronic conduction network established by the electron tunnelling mechanism tends to be perfect, and the continuous increase of conductive fillers will not have distinct effects on the improvement of the electrical conductivity.

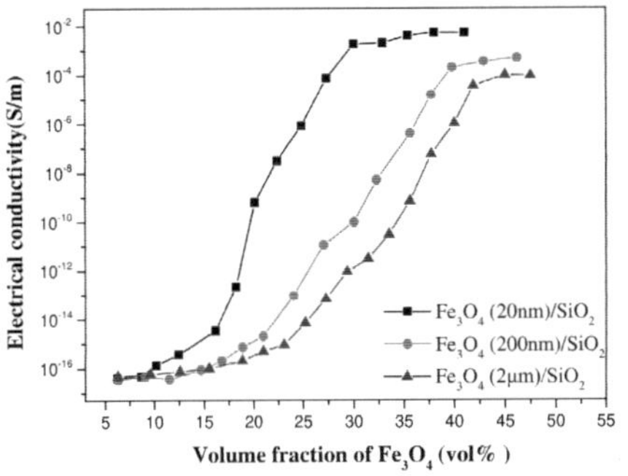

Figure 2. The effect of conductive particle size on the electrical conductivity of composites.

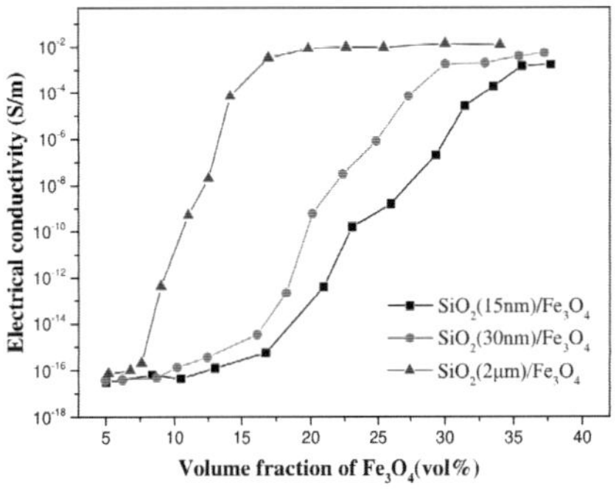

Figure 3. The effect of matrix particle size on the electrical conductivity of composites.

3.2. Electrical conductivity of composites

The effect of the particle size on the electrical conductivity of composites simultaneously consisted of Fe_3O_4 and SiO_2 was not fully investigate. Moreover, it is difficult to achieve particles with identical sizes. Under these circumstances, the influences of particle size on electrical conductivity are analysed by means of 3D simulation model.

The trend of the electrical conductivity of composites shown in figure 2 is obviously related to conductive particle size. Although the composites consisting of small-sized conductive particles present higher electrical conductivity under low volume fraction, the conductivity tends to change more slowly with the volume fraction and the maximum conductivity is lower than that of composites consisting of large-sized conductive particles. The series-parallel channels based on tunnelling theory are sensitive to the distance between the particles and the number of particles involved in the establishment of the conductive paths. The average distance between small-sized particles is large, which makes it difficult to form conductive paths between conductive particles. However, the number of small-sized particles participating in the formation of conductive paths increases faster than that of large-sized particles. It can be concluded that the number of conductive paths generated in composites with small-sized conductive particles is naturally larger than that in composites with large-sized conductive particles.

Additionally, the effects of different matrix particle sizes on electrical conductivity are studied. It can be seen from figure 3 that the electrical conductivity of the composite based on micron-sized SiO_2 is obviously higher than that of the composite based on nano-sized SiO_2. The difference in electrical properties of Fe_3O_4/SiO_2 composites with varying size SiO_2 as matrix is attributed to the effect of matrix size on electron tunnelling and the establishment of conductive paths. It is obvious that the pores of micron-sized SiO_2 matrix are bigger than those in nano-sized SiO_2 matrix. Therefore, the conductive nanoparticles distributed in the pores of micron-sized matrix will be more easily concentrated, so the probability of conductive particles conducting with each other is higher, hence the electrical conductivity will show a higher level. Furthermore, since nano-sized SiO_2 is closer to the size of fillers, nano-sized matrix particles evenly distributing around the fillers act as a "wall" that prevents the electron transfer.

4. Conclusions

By establishing the 3D simulation model of inorganic conductive network, the conductive mechanisms and the electrical conductivity of Fe_3O_4/SiO_2 composites are studied. The electrical conductivity of composites is found positively relevant to conductive particle content. Micron-sized matrix particles are more conducive to the electrical conductivity of composites, and the electrical conductivity of composites based on micron-sized SiO_2 can reach a higher level by adding a small quantity of conductive fillers. On the contrary, small size conductive particles contribute to improving the conductivity of composites. This non-liner trend of electrical conductivity is consistent with the percolation theory, which further verifies that the established model is effective. The conductive mechanism of inorganic conductive network is the basis of the further study on organic conductive network in sediments, and the 3D simulation model of inorganic conductive network proposed in this work also provides a fresh idea for further study on long-distance electron transport and biogeochemical process in sediments.

Acknowledgments

This work was financially supported by the National Natural Science Foundation of China (51879042, 51839011); Postgraduate Research & Practice Innovation Program of Jiangsu Province, the Fundamental Research Funds for the Central Universities (KYCX20_0105); State Key Laboratory of Nuclear Power Safety Monitoring Technology and Equipment (K-A2020.415); National Key Research and Development Program of China (2018YFE0125500).

5. References

[1] Nielsen LP, Risgaard-Petersen N, Fossing H, Christensen PB and Sayama M 2010 *Nature.* **463**(25) 1071-4
[2] Hermans M, Lenstra W K, Hidalgo-Martinez S, Helmond N V, Witbaard R, Meysman F, Gonzalez S and Slomp C P 2019 *Environmental Science & Technology.* **53**(13) 7494-503
[3] Bo B J, Findlay A J, Pellerin A 2019 *Frontiers in Microbiology.* **24**(10) 849
[4] Kato S, Hashimoto K, Watanabe K 2012 *Proceedings of the National Academy of Sciences.* **109**(25) 10042-6
[5] Malvankar N S, King G M and Lovley D R 2015 *The ISME Journal.* **9**(2) 527-31
[6] Liu F, Rotaru A, Shrestha P, Malvankar N, Nevin K and Lovley D 2015 *Environmental Microbiology.* **17**(3) 648-55
[7] Shinde Mahesh A and Kim Haekyoung 2021 *Materials Today Communications.* **26**
[8] Judd J 2004 *Fire Prevention & Fire Engineers Journal.* **381** 36
[9] Simmons J G 1963 *Journal of Applied Physics.* **34**(9) 2581-90
[10] Irzhak V I 2021 *Colloid Journal.* **83**(1) 64-9
[11] Ying H, Wang W, Xiao Z, Guo X, and Zhang Y 2018 *Journal of Applied Polymer Science.* **135**(32) 46517
[12] Wang G, Wang C, Zhang F and Yu X 2018 *Computational Materials Science.* **150** 102-6
[13] Guanda Yang, Dirk W Schubert, Fritjof Nilsson, Muchao Qu and Michael Redel 2020 *Macromolecular Theory and Simulations.* **29**(5)
[14] Balberg I 2009 *Journal of Physics D Applied Physics.* **42**(6) 64003-18
[15] Chen X, Alian A R and Meguid S A 2020 *European Journal of Mechanics - A/Solids.* **84** 104053
[16] Simmons J G 1963 *Journal of Applied Physics.* **34**(6) 1793-803
[17] Dongjae Kim and Jaewook Nam 2020 *Phys. Chem. C.* **124**(1) 986-96
[18] Ning H, Karube Y, Cheng Y, Masuda Z and Fukunaga H 2008 *Acta Materialia.* **56**(13) 2929-36
[19] Milind Jagota and Isaac Scheinfeld 2020 *Phys. Rev. E.* **101** 012304

Investigation of Base Transport Mechanism in Silicon-Germanium Heterojunction Bipolar Transistor Operating over Wide Temperature Range

Guofang Yu, Jie Cui, Yue Zhao, Jun Fu*, Tian-Ling Ren

School of Integrated Circuits & Beijing National Research Center for Information Science and Technology (BNRist), Tsinghua University, Beijing 100084, China

fujun@tsinghua.edu.cn

Abstract. A high breakdown voltage silicon-germanium heterojunction bipolar transistor operated over a wide temperature range from 300 K to 10 K has been investigated. The measured Gummel characteristics illustrate that the collector current and base current both shift to the higher voltage as the temperature decreases. The f_T/f_{max} are extracted to be 23/40 GHz at 300K, 28/40 GHz at 90 K, and 25/37GHz at 10K, respectively. The effective amplification range becomes narrow as the temperature decreases. And the ideality factor of base current in the low current region is shown to be temperature-dependent and its value is much larger than 2 at cryogenic temperatures. This phenomenon indicates that the base current is not only contributed by drift, diffusion, and Shockley-Read-Hall recombination, but also by trap-assisted tunneling. The Hurkx local trap-assisted tunneling has been used to analyze the non-ideal base transport mechanism. And a calibrated TCAD device model is developed to further verify this non-ideal transport mechanism.

1. Introduction

Due to the excellent current gain, radio frequency (RF), and noise characteristics at cryogenic temperature, silicon-germanium heterojunction bipolar transistor (SiGe HBT) has attracted significant attention for quantum computing [1], [2]. For instance, SiGe HBT can be used as a pre-amplifier in quantum computing readout circuit, where the extremely low temperature drops down to 4.2 K, even down to tens of mK [3]. It is important to understand the cryogenic transport mechanisms of SiGe HBT for successful cryogenic applications. At cryogenic temperature, quantum transport mechanisms such as trap-assisted tunneling (TAT) and direct tunneling become non-negligible [4]. John D. Cressler et al have focused on studying the tunneling mechanism of collector current, and found that the direct tunneling was sensitive to the base width and Ge profile [5], [6]. Base transport mechanism has also been investigated, and an analytic TAT model of base current has been developed to describe the non-ideal characteristics [7]. The excess base current caused by non-ideal transport is one of the critical factors affecting the effective amplification range at low temperatures. The excess current leads to a large quiescent operating point, which causes high quiescent power dissipation. This power dissipation will be converted into heat, resulting in the variation of the low-temperature environment. Due to the quantum information is sensitive to the change of temperature, it is important to decrease the quiescent operating point to meet the low power dissipation requirement. However, the base non-ideal transport mechanism affecting the effective operating point at low temperatures has not been studied adequately. In this work, the direct current (DC) and RF performances of SiGe HBT are characterized from 300 K down to 10K. The TAT effect in base-emitter junction is introduced to analyze the non-ideal transport mechanism of base current in the low current region. And TCAD simulation calibrated by the device structure and doping profile of the investigated SiGe-HBT has been taken to verify the transport mechanism.

2. Device and Experimental Details

The investigated NPN SiGe HBT device was fabricated by the 0.5 μm SiGe technology and designed as a CEBEC layout structure with an emitter geometry of 2 × (0.6 × 20) μm^2 and a high BV_{CEO} of 6 V at room temperature. The cross-section schematic of the investigated SiGe HBT is shown in Fig.1. After the formation of an n+ sub-collector on the p-type substrate, the lightly phosphorus-doped n− collector epitaxy was deposited. An ultra-thin SiGe base layer with heavy boron doping was epitaxy grown with 14% Ge content. And the device was fabricated with a lightly selectively implanted collector (SIC), a heavily doped poly-Si emitter, and a heavy boron-doped extrinsic polysilicon base. Local oxidation of silicon (LOCOS) was used for device isolation.

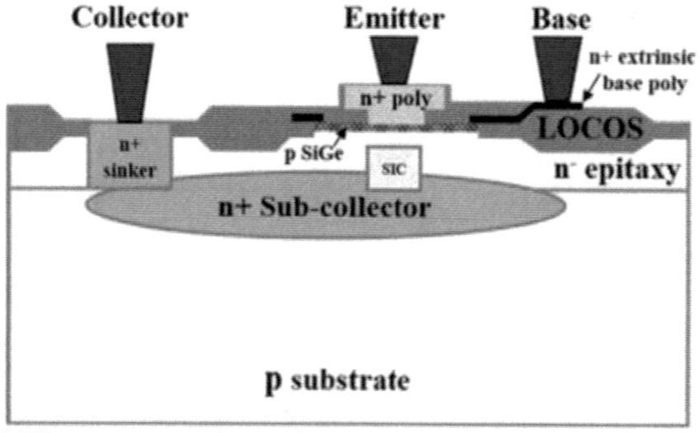

Figure.1 The cross-section schematic of the investigated SiGe HBT.

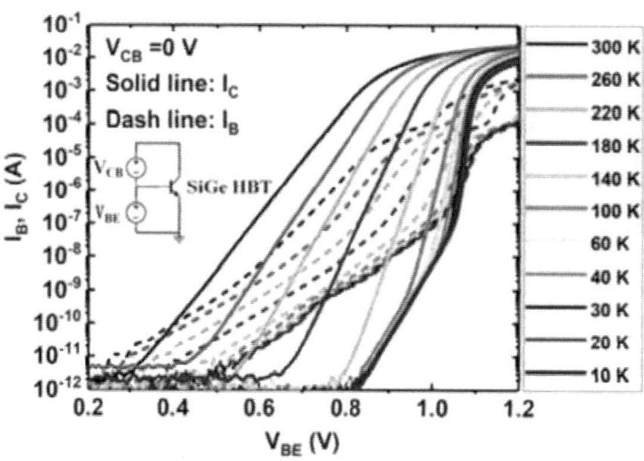

Figure.2 Forward Gummel characteristics of measured SiGe HBT from 300 K to 10 K.

Device measurements were performed over the wide temperature range from 300 K down to 10 K, which was provided by the Lakeshore Cryogenic Probe Station. To characterize the SiGe HBT's electrical behavior across temperature, the Gummel characteristics were measured by Agilent B1500A Semiconductor Parameter Analyzer and the S parameters were measured by Keysight PNA-X Network Analyzer from 300 K to 10 K. The forward-active mode Gummel characteristics at V_{CB} = 0 V from 300 K to 10 K are shown in Fig.2. These measurement results illustrate that the collector current (I_C) and base current (I_B) both shift to higher operating voltage as the temperature decreases. After being de-embedded with open and short test structures, the f_T/f_{max} are extracted from S parameters at V_{CE} = 2 V, as shown in Fig.3. The f_T/f_{max} have a value of 23/40 GHz at 300K, 28/40 GHz at 90 K, and 25/37GHz at 10K, respectively.

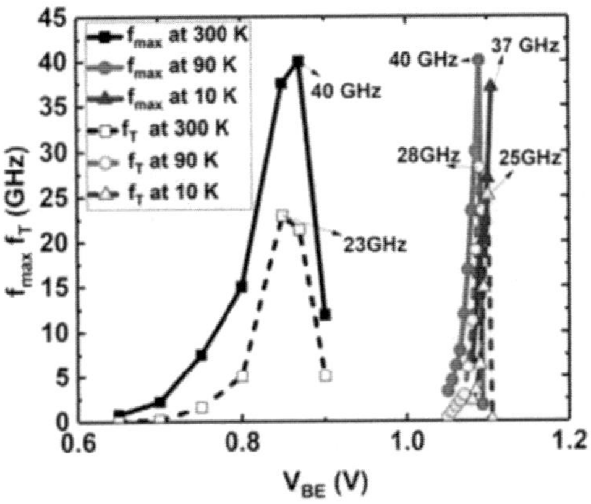

Figure.3 The f_T/f_{max} extracted from S parameters at 300 K, 90 K and 10 K.

3. Trap-assisted Tunneling Analysis

As shown in Fig.2, the forward-active mode Gummel characteristics of SiGe HBT were measured at $V_{CB} = 0$ V, which could avoid the additional effects between base and collector caused by junction bias. The non-ideal current in the low current region, especially for base current, becomes more and more obvious when the temperature decreases. The current gain attenuation is caused by the excess base current in the low current region, consequently resulting in the narrowing of the effective amplification range. Fig.4 shows the current gain ($\beta = I_c/I_B$) varying with I_c at different temperatures. The inset of Fig.4 presents the full width at half maxima (FWHM) of current gain as FWHM = $\log(I_{F.\max}) - \log(I_{F.min})$, where $I_{F.min}$ and $I_{F.max}$ are the minimum and maximum collector current at half peak current gain respectively. The results show that the effective amplification range of SiGe HBT becomes more and more narrow as temperature goes down to 10 K. The extracted FWHM and $I_{F.min}$ are plotted in Fig.5. It is obvious that the FWHM decreases, but the $I_{F.min}$ increases with the temperature cooling. The excess base current caused by the non-ideal transport mechanism at the base-emitter junction is one of the most important factors contributing to the decreasing of FWHM and the increasing of $I_{F.min}$.

Fig.6 shows the ideality factor of base current in the low current region ($N_{F,I_{BL}}$) extracted from Gummel characteristics. The extraction method is shown in the inset of Fig.6, where k is the Boltzmann constant, q is the electron charge and T is the ambient temperature. The $N_{F,I_{BL}}$ is equal to 2 at 300 K and increases to 56 as the temperature is lowered to 10 K. The recombination theory of Shockley-Read-Hall (SRH) has illustrated that the ideality factor must be among 1 to 2. However, there are a large number of traps existing in the base-emitter space charge region (BE SCR). The base current can be aided by Hurkx trap-assisted tunneling (TAT) [4].

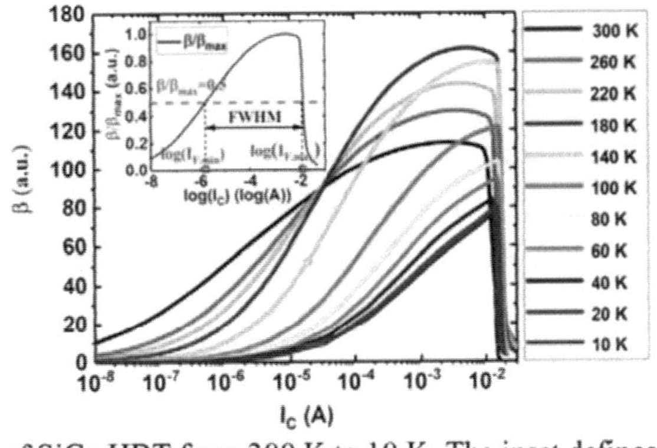

Figure.4 Current gain of SiGe HBT from 300 K to 10 K. The inset defines the FWHM of current gain.

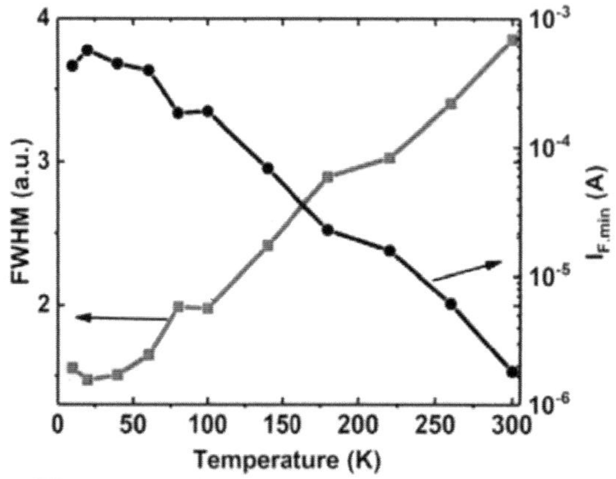

Figure.5 The extracted FWHM and $I_{F.min}$ from 300 K to 10 K.

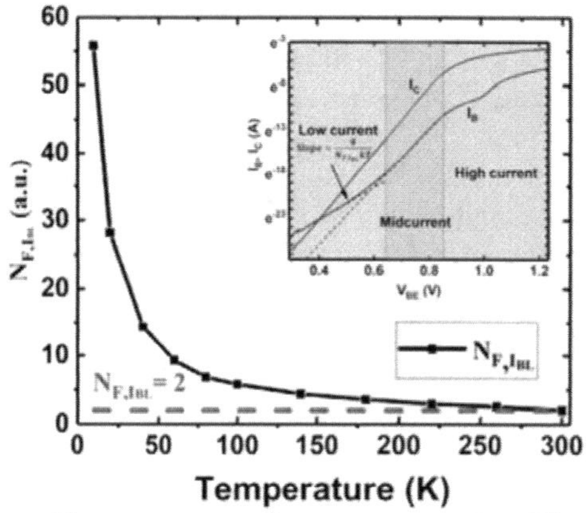

Figure.6 The ideality factor of base current at low current region. The inset shows the extraction method.

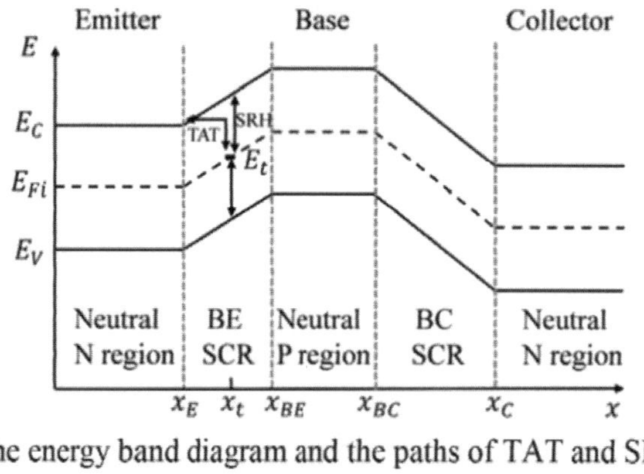

Figure.7 The energy band diagram and the paths of TAT and SRH process.

The energy band diagram and the paths of the TAT and SRH process are shown in Fig.7. For the conventional SRH theory, the recombination center E_t captures and emits carriers at the location x_t. However, due to the heavy doping of both emitter and base region which leads to a narrow BE SCR, an electron in conduction band at location $x < x_t$ may have a probability of tunneling to a trap E_t at location x_t, and have a chance of being captured [8]. Because of the narrow BE SCR and a large number of traps inside the bandgap, the excess base current caused by TAT cannot be neglected, especially at low temperatures. It is well known that the probability of tunneling is weak temperature-dependent. Hence, the base current in the low current region is dominated by TAT current and the slope is almost constant with temperature varying, as shown in Fig.2.

4. TCAD Simulation and discussion

A 2-D device model is created in Sentaurus TCAD to focus on studying the TAT mechanism in base current over the wide temperature range. The device structure and doping profile of the measured SiGe HBTs are used to calibrate the simulation device model. Simulations are performed by using the hydrodynamic transport model, Phillips unified mobility model, Slotboom bandgap narrowing model, and the Okuto-Crowell avalanche model to match the DC characteristics at 300 K. The local Hurkx model is used to simulate the TAT current of base current.

For the conventional SRH theory, the recombination rate can be expressed as following [8]:

$$\mathrm{SRH} = \frac{n_i^2\left(\exp\left(\frac{qV_{BE}}{kT}\right)-1\right)}{\tau_p(n+n_l)+\tau_n(p+p_l)} \quad (1)$$

Where $n_l = n_i \exp\left(\frac{E_t-E_i}{kT}\right)$ and $p_l = n_i \exp\left(\frac{E_i-E_t}{kT}\right)$, n_i is the intrinsic carrier density, E_i and E_t denote intrinsic Femi energy and trap level respectively, n (p) is electron (hole) density, $\tau_{n(p)}$ is electron (hole) SRH lifetime. It has been clear that the ideality factor of SRH recombination current cannot be larger than 2. But when the Hurkx model is considered, the carrier lifetime is:

$$\tau_{n(p)}^{TAT} = \frac{\tau_{n(p)}}{1+\Gamma_{TAT}} \quad (2)$$

Where, Γ_{TAT} is the TAT factor which is related to trapping level and the probability of tunneling. By replacing the $\tau_{n(p)}$ by $\tau_{n(p)}^{TAT}$, we can obtain the TAT recombination $\mathrm{SRH}^{\mathrm{TAT}}$. The TAT recombination current can be modeled by:

$$\mathrm{I} = \mathrm{qA}\int_{x_{BE}}^{x_E} \mathrm{SRH}^{TAT}\, dx \quad (3)$$

Where A is the area of emitter.

Comparison of I_C and I_B between experiment and simulation results using Hurkx TAT model at 300 K, 260 K, and 140 K is shown in Fig.8. The excess base current in the low current region can be well described by the TAT model, which is presented in Eq. (3). Since the heterojunction barrier effects are not considered during the simulation, there is an existing deviation of the base current in the high current region between simulation and experiment [9]. When the temperature is lowered, the collector current and mid base current are higher than the experiment results. This phenomenon may be caused by the temperature dependence of bandgap narrowing introduced by the heavy doping [10].

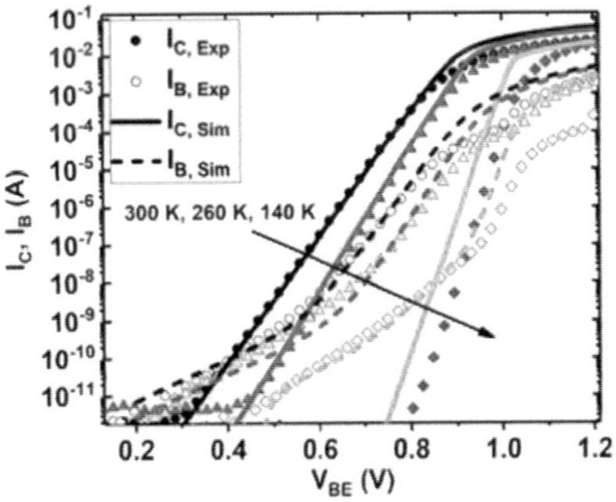

Fig.8 Comparison of collector current and base current between experiment and simulation results using Hurkx TAT model at 300 K, 260K and 140 K

5. Conclusion

A high breakdown voltage SiGe HBT operated over the wide temperature range from 300 K to 10 K has been investigated. The measured DC and RF characteristics have illustrated that the collector current and base current both shift to the higher voltage as the temperature decreases. The excess base current caused by the non-ideal transport mechanism is one of the dominant factors leading to the narrowing of the effective amplification range. The ideality factor of base current in low current region is shown to be temperature-dependent and much larger than 2 at low temperatures, indicating that the current is not only contributed by drift, diffusion, and Shockley-Read-Hall recombination, but also by TAT. The Hurkx local trap-assisted tunneling model has been used to analyze this non-ideal base transport mechanism. And a calibrated TCAD device model is simulated to verify this non-ideal transport mechanism. The simulation results show that the TAT mechanism dominates the excess base current at low temperatures.

Acknowledgment

This work was supported by the National Key Research and Development Program (2016YFA0200400), National Natural Science Foundation (61434001, 61574083, 92064002), and National Basic Research Program (2015CB352101) of China. The authors are also thankful for the support of the Research Fund from Beijing Innovation Center for Future Chip, Beijing Natural Science Foundation (4184091), and Shenzhen Science and Technology Program (JCYJ2015083119222414 6).

6. References

[1] W.-T. Wong, M. Hosseini, H. Rucker, and J. C. Bardin, "A 1 mW Cryogenic LNA Exploiting Optimized SiGe HBTs to Achieve an Average Noise Temperature of 3.2 K from 4–8 GHz," IEEE/MTT-S International Microwave Symposium (IMS), Los Angeles, CA, USA, Aug. 2020, pp. 181–184, doi: 10.1109/IMS30576.2020.9224049.

[2] P. S. Chakraborty, A. S. Cardoso, B. R. Wier, A. P. Omprakash, J. D. Cressler, *et al.*, "A 0.8 THz fmax SiGe HBT Operating at 4.3 K," IEEE Electron Device Lett., vol. 35, no. 2, pp. 151–153, Feb. 2014, doi: 10.1109/LED.2013.2295214.

[3] M. J. Curry, T. D. England, N. C. Bishop, G. Ten-Eyck, J. R. Wendt *et al.*, "Cryogenic preamplification of a single-electron-transistor using a silicon-germanium heterojunction-bipolar-transistor," Appl. Phys. Lett., vol. 106, no. 20, p. 203505, May 2015, doi: 10.1063/1.4921308.

[4] G. A. M. Hurkx, D. B. M. Klaassen, M. P. G. Knuvers, and F. G. O'Hara, "A new recombination model describing heavy-doping effects and low-temperature behaviour," International Technical Digest on Electron Devices Meeting, Washington, DC, USA, 1989, pp. 307–310, doi: 10.1109/IEDM.1989.74285.

[5] H. Ying , J. Dark, A. P. Omprakash , B. R. Wier , L. Ge *et al.*, "Collector Transport in SiGe HBTs Operating at Cryogenic Temperatures," IEEE Transactions on Electron Devices, vol. 65, no. 9, pp. 3697–3703, Sep. 2018, doi: 10.1109/TED.2018.2854288.

[6] D. Davidović, H. Ying, J. Dark, B. R. Wier, L. Ge *et al.*, "Tunneling, Current Gain, and Transconductance in Silicon-Germanium Heterojunction Bipolar Transistors Operating at Millikelvin Temperatures," Phys. Rev. Applied, vol. 8, no. 2, p. 024015, Aug. 2017, doi: 10.1103/PhysRevApplied.8.024015.

[7] Z. Xu, G. Niu, L. Luo, and J. Cressler, "A Physics-Based Trap-Assisted Tunneling Current Model for Cryogenic Temperature Compact Modeling of SiGe HBTs," ECS Trans., vol. 33, no. 6, pp. 301–310, Dec. 2019, doi: 10.1149/1.3487560.

[8] G. A. M. Hurkx, D. B. M. Klaassen, and M. P. G. Knuvers, "A new recombination model for device simulation including tunneling," IEEE Trans. Electron Devices, vol. 39, no. 2, pp. 331–338, Feb. 1992, doi: 10.1109/16.121690.

[9] K. Lee, D. Cho, K. Park, and B. Kim, "Improved VBIC model for SiGe HBTs with an unified model of heterojunction barrier effects," IEEE Trans. Electron Devices, vol. 53, no. 4, pp. 743–752, Apr. 2006, doi: 10.1109/TED.2006.871194.

[10] S. E. Swirhun, D. E. Kane, and R. M. Swanson, "Temperature dependence of minority electron mobility and bandgap narrowing in p+ Si," Technical Digest., International Electron Devices Meeting, San Francisco, CA, USA, 1988, pp. 298–301, doi: 10.1109/IEDM.1988.32816.

Research on Design Electrode Channel in Microfluidic Chip

Zhou yingmei[1,*], Lv Jian[2]

[1]Printing and packaging Department, Shanghai Publishing and Printing College, Shanghai, China

[2]Publishing and Printing College, University of Shanghai for Science and Technology, Shanghai, China

47131532@qq.com

Abstract. In recent years, microfluidic chips have been widely used in food safety, environment detection, clinical diagnosis and biochemical. However, Digital microfluidic chips(DMF) are attracting more and more eyes because picoliter-microliter-sized sample droplets can be accurately and automatically manipulated. In this paper, a method of design electrode channel height is studied to improve the efficiency of the voltage driving.

1. Introduction

As the development of MEMS technology and the interinfiltration between different subjects, the Lab-on-a-chip(LOC), which realizes the full auto-process and auto-analysis on a single chip, has attracted our attention in the future. Electrowetting-on-dielectric(EWOD)based digital microfluidics has been widely searched due to its unique advantages such as simple structure, simple operation and strong capability[1]. The theory of EWOD is shown in Fig.1, which obviously is accepted by us with the structure. The channel height between electrodes is an important part. Traditional chip still has low separation efficiency, high voltage, and poor accuracy, which limited the application and development. The digital microfluidic chips include electrode, dielectric material, hydrophobic layer and droplets. The factors affecting the performance of chips are above these. However, the channel height(d) and electrode size(L) are related to the voltage. According to driving theory, the contact angle hysteresis phenomenon required increased voltage value to drive the drops. So the advanced forum occurred. The separate voltage limit forum is about changing into like this in (1). The voltage must increase with the result of d/L with other conditions unchanged[2]. The material of dielectric layer was also related to the voltage. And the material of dielectric constant can be selected to reduce the voltage.

Journal of Physics: Conference Series **2065** (2021) 012014 doi:10.1088/1742-6596/2065/1/012014

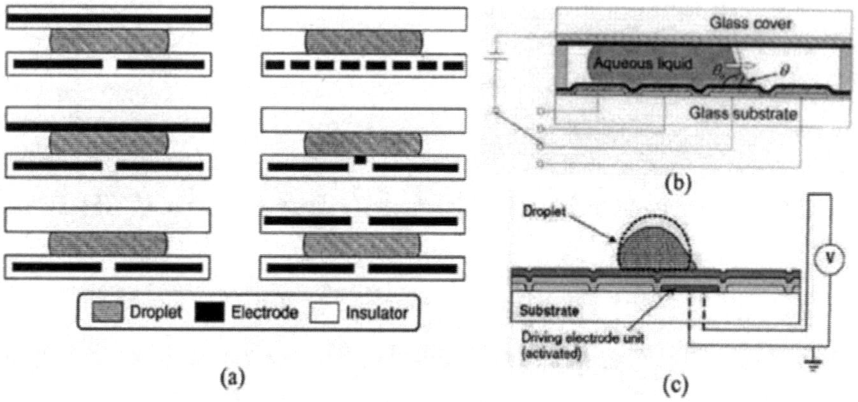

Figure 1. Structure of EWOD Chips

$$v^2 - v_T^2 = \left(\frac{1}{R_2} - \frac{1}{R_1}\right)\frac{2\,dt \quad \gamma l_\delta}{\cos a\varepsilon_r\varepsilon_0} \qquad (1)$$

In (1-1), γl_δ means the surface tension, V means the electric potential, VT coms from the threshold Voltage, D is the channel height,R1 and R2 are the radius of curvature,ε_0 *is* dielectric constant of vacuum,ε_r means relative dielectric constant of dielectric layer,α is on behalf of surface angle, t equals to the thickness of dielectric layer. Suppose R2≈L/2, R1≈L (N*²+1) /2, so

$$V^2 - v_T^2 \approx \left(1 - \frac{1}{N^{*2}+1}\right)\frac{4\,dt \quad \gamma l_\delta}{L\cos a\varepsilon_r\varepsilon_0} \qquad (2)$$

The N* means the numbers of electrode which are divided from one drop,α and $\gamma l_\delta$are related with the environment, the only reason influenced voltage is the value of d/L.

According to the former different electrode shape design, the triangle interdigitated finger electrode was better than the rectangular interdigitated finger electrode design effect in Fig.2[3].It showed higher voltage driving ability and driving speed. From another way, we can get the result that the sharper corner of triangles has, the smaller gap between electrodes. Even though there are various types electrode edge design, the practical application design mainly focuses on the normal electrode shape, such as rectangular electrode. How to improve the rectangle electrode voltage driving ability and speed? How to avoid the influence coming from the channel height?

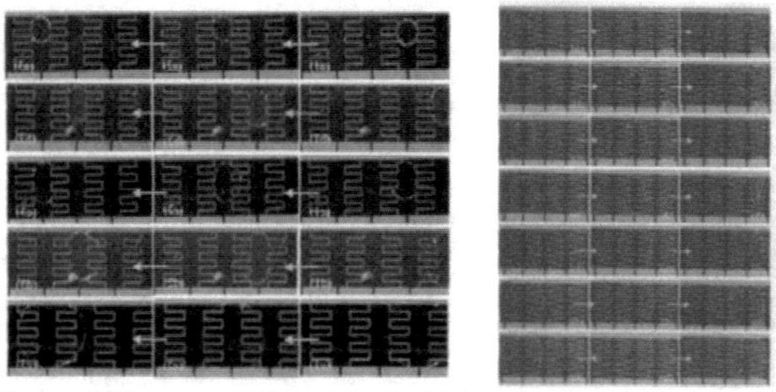

Figure 2. Different Volumes of Droplets in Triangle and Rectangle Electrode

Many research and patent did studies on this filed[4]. The rectangle steps produced through making process in Figure.3(left) The gap is so big that the droplet moving efficient got influence. With the development of manufacturing process in EWOD, the dielectrode film mulching in Figure.3(right) is used to reduce the steps of channel height to make it smooth. So that the driving speed and ability are better than the left method.

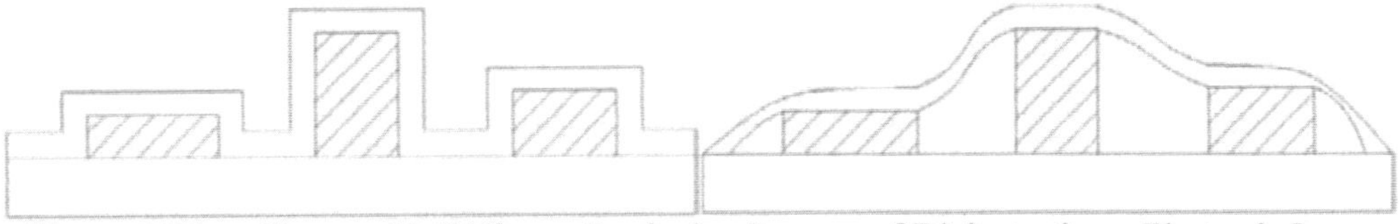

Figure.3 Wave Shape with Different Producing Process of Dielectrode on Electrode Layer

In this paper, the author wants to try a novel way to change the producing process, which maybe simple the technology, reduce the cost and get high benefit.

2. Experiment

Most EWOD chips set the dielectrode layer on the electrode layer in Fig.4[5]. So the channel height will produce following the electrode inks thickness[6].

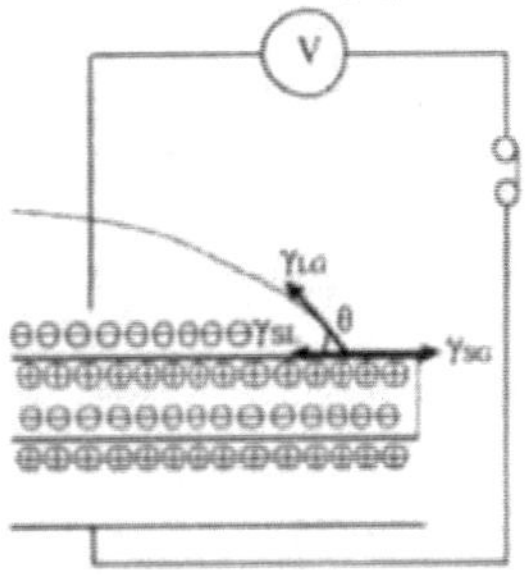

Figure 4. The Model of EWOD Theory

According to the original thinking, normal and complex pattern are designed with the software Adobe Illustrator. There are same channel sizes among the normal rectangle electrodes. The distance is about 0.3mm, the electrodes sizes are about 2-4mm, so the microscope image pattern is very visible in Fig.5. In order to consider different kinds of designs as totally as possible, the novel pattern electrodes are designed in Fig.5 (right). The pictures are used to test the dieletrode layer uniform. The electrodes edge has narrow or wide gap. They look so complex that it is difficult to set the guide line from the electrode.

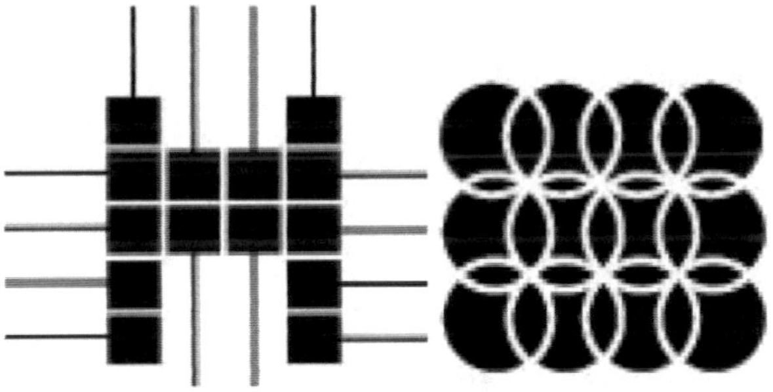

Figure 5. Design pattern of Electrodes

The purpose of next process is to output these patterns on the basement layer. To keep the dielectrode layer uniform, printing the designed patterns on the prepared dielectrode layer is a creative method to solve these questions such as gap size effect, driving speed and driving ability. For complex pattern, another way to set guide line to power from electrode edge is to design a new layer under electrode layer which avoid circuit shortcut in Fig.6. The top layer means hydrophobic layer, the second layer means dielectrode layer, the bottom layer three rectangles mean electrodes.

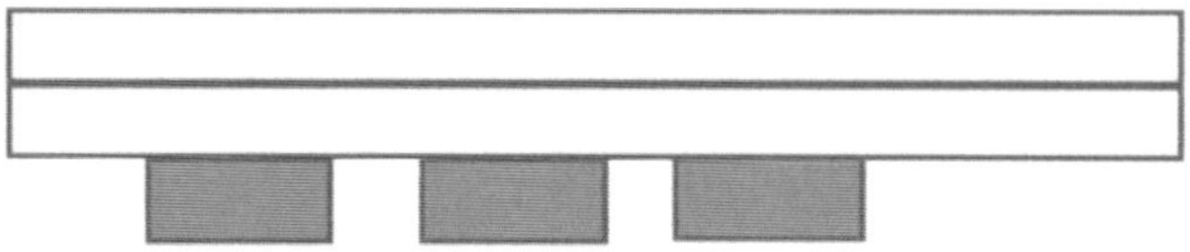

Figure 6. Relationship with Hydrophobic, Dielectrode and Electrode Layer

3. Analysis and Conclusion

In this paper, the experiment was printed with Inkjet printer (Dimatix 2831) for test. With different purposes design, the electrodes printed into PI film such as three layers in Figure 7.

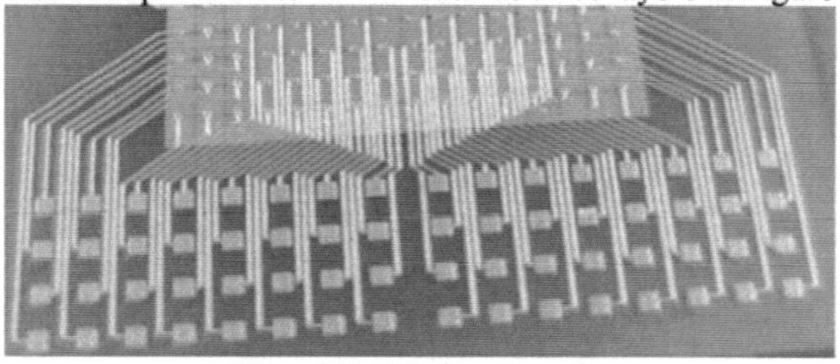

Figure 7. Printing Sample

In order to verify the effect on different between droplet sizes, various droplets sizes should be used as Dimeter A=0.6mm, B=0.8mm, C=1.00mm, D=1.20mm and E=1.40mm with gap size as 0.3mm and 0.6mm to test the driving speed in Figure.8.

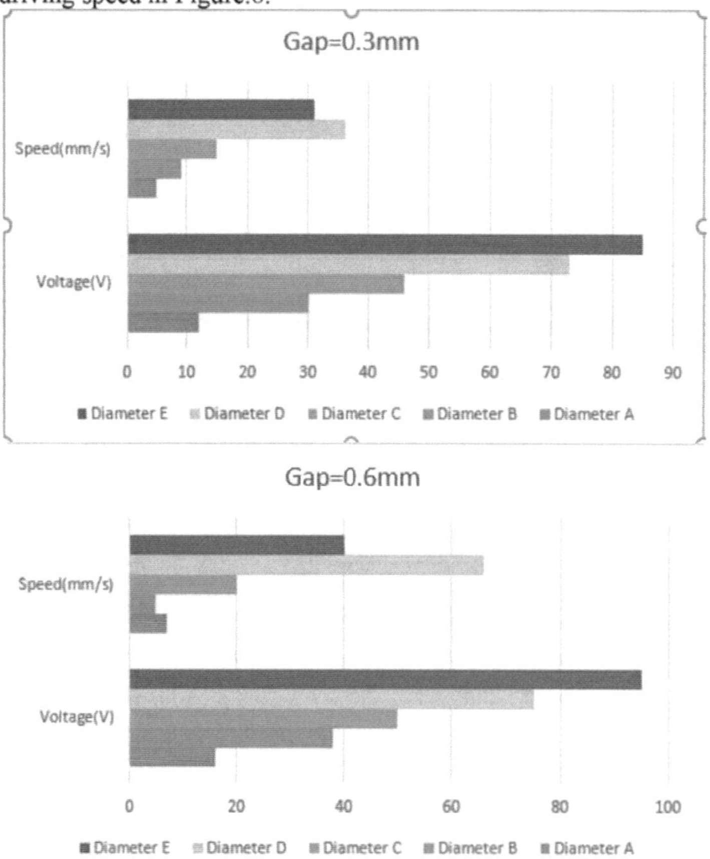

Figure 8. Comparison of Different Channel Gap Size According to Various Droplets

From the Fig.7 analysis, we found that new way to print electrode on the back side of dielectrode completely missed the step influence. Then the channel gap couldn't take any effect on driving ability. The data showed that driving speed is suitable to the voltage variable. The droplets driving speed isn't linear to the gap size change.

This paper only compare limited filed to prove the channel size influence, offer one way to set electrode on the back of dielectrode layer because of search time. In fact, we have more ways to compare to solve the possible question. One is to fill the gap between electrodes with normal ink to ensure the surface smooth. The other way is to think about the normal ink layer with hole to link the electrode and guide line to power. We will discuss in the future.

The purpose in this paper to try the novel way is to use better produce process to improve the electrode accuracy and channel gap, so the droplets size will be researched and developed to meet the volume of the business requirement.

Acknowledgements

This paper is supported by Key Lab of Intelligent and Green Flexographic Printing: Research on Quality Control Strip by Flexography Printing(ZBKT201901).

4. Reference

[1] Zhou Yi-hua, *Application of Inkjet Printing in Microfluidic Paper-Based Analytical Device*, Digital Printing, Vol203(2019)1-6.

[2] Yiqiang Fan. *Low-cost and Flexible Film-based Digital Microfluidic Device*[J]. Micro and Nano Letters,2020 Vol.15 No.3 P163-165 1750-0433

[3] Dongping Chai, *Research on Droplet Transport Method of Digital Microfluidic Chip Based on Screen Printing*[M], 2020, Beijing

[4] Zhaolong wang, *Research on Digital Microfluidic Devices for High Performance Droplet Separation Based on Electrowetting on dielectric*[M], 2020

[5] Zehui Gang, *Driver System Design of Digital Microfluidic Chip*[M], 2019

[6] Jianfeng Chen, *Study of Novel Digital Microfluidic Chips Based on EWOD*[M],2014

Journal of Physics: Conference Series **2065** (2021) 012015 doi:10.1088/1742-6596/2065/1/012015

Wafer Direct Technology for Mini LED Flip Attachment

Kazuo Hiraki

Suzuki Co., Ltd. 2014-5, Ogawara, Suzaka Nagano 382-8588 Japan

hiraki_kazuo@suzukinet.co.jp

Abstract. We would like to introduce the mechanism structure of our wafer direct die attachment technology for mini LED. Wafer direct technology is structured from two parts of that one is adhesive one shot dipping and other is pinhole free chip shooting. 1st, we developed "One shot wafer dipping" which can transfer the adhesive to around 30,000 chips of mini LED on a wafer at once. Second section is to take about die mounting. Common die bonder use an ejector pin to pick up mini LED chip. Ejector pin makes a pinhole on wafer and lead air into between wafer and mini LED chip. This air is effective to remove mini LED chip from wafer by the pick up nozzle. Ejector pin of die bonder leaves a broken piece of wafer on chip. We developed wafer direct chip shooting technology. Wafer direct chip shooter also use an ejector pin for shooting down mini LED chip. But wafer direct does not make any pinhole on wafer that means no broken piece of wafer exists. Our wafer direct technology maximize productivity of mini LED die bonding and minimize faulty ratio of mini LED die bonding.

1.Introduction [1] Wafer direct dipping.

We considered to sandwich a wafer between pusher head and dipping stage. See figure 1 "method of wafer dipping", 1. Squeegeeing of ACP for flattening and control thickness. 2. ACP is flatten in targeted thickness. 3. Dipping stage comes to bottom side of wafer. 4. Pusher head comes down and dipping stage goes up. 5. ACP is transferred to mini LED chips on wafer by one shot. We use a round plate for the pusher head as like a wafer expander. Dipping stage rise 5mm up and pusher head push 1mm down a wafer for dipping. We set the matching point of wafer and ACP as zero height level. We keep 3 seconds dipping time at zero height level to transfer ACP to wafer by surface tension. Then we keep pusher head at zero height level and dipping stage comes down in slow speed of [0.01mm per second] for optimization of dipping volume of ACP. It takes 60 seconds as one dipping cycle that is too slow speed for common LED production. (See figure 2. "Timing chart of dipping"). But more than 30 thousands mini LED chips are existing on single wafer which means to take only 0.002 seconds per chip. So wafer direct dipping can meet to huge demand of mini LED bonding. It is easy to optimize supply volume of ACP or any adhesive in wafer direct dipping. Wafer direct dipping does not occur any offset of ACP or adhesive supply position as like often happen on solder printing. We use stretching of wafer for making flatness. But we have to keep return tension ability of wafer until next chip mounting process. So we minimize a distance of extension to push a wafer down for dipping. Minimum extension of wafer for flattening is 1mm.

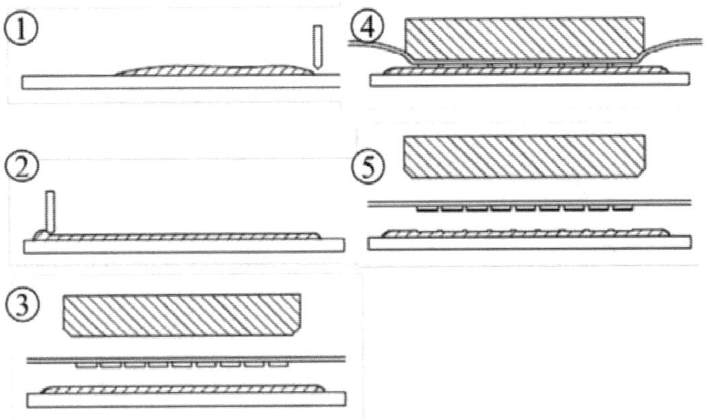

Figure.1. Method of wafer dipping

Wafer direct one shot dipping can transfer 60% depth of ACP pool to mini LED chips on wafer by surface tension. We can control dipping volume by depth of ACP pool and dipping time profile. See Figure. 2 dipping volume control and Figure. 3 time chart of dipping. We can control depth of ACP pool from 10μm up to 500μm by squeegeeing. We can also control dipping volume by shorten or extend of waiting time and separating speed of wafer and ACP pool.

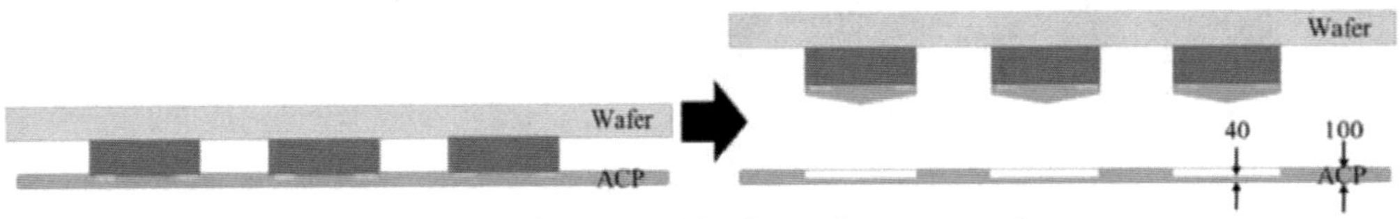

Figure. 2 Dipping volume control

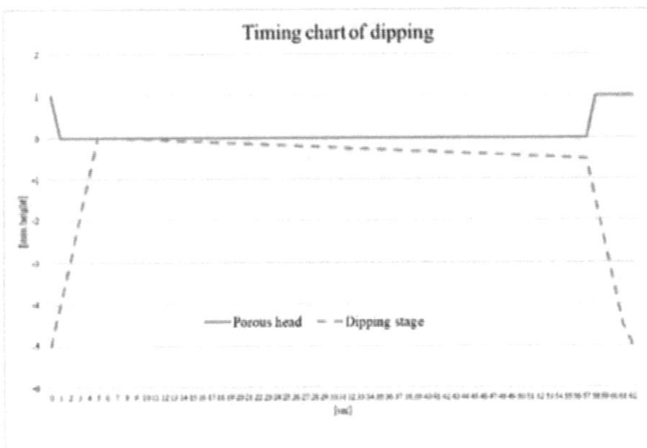

Figure.3 Timing chart of dipping

2. Consideration and result [1] Wafer direct dipping.

We prepared a drilled pusher head with 380 mini vacuum holes which has for flattening of wafer. Unfortunately, drilled head couldn't get complete flatness of wafer which couldn't transfer ACP to half of 30,000 mini LED chips on wafer. We checked dipping condition of mini LED chips and found that dipping of mini LED chips at around vacuum hole area are done well. But mini LED chips at far from vacuum hole area are not stable condition. We found that many airspace existed between vacuum hole and vacuum hole which disturbed flatness of wafer. See Fig. 4 Airspace.

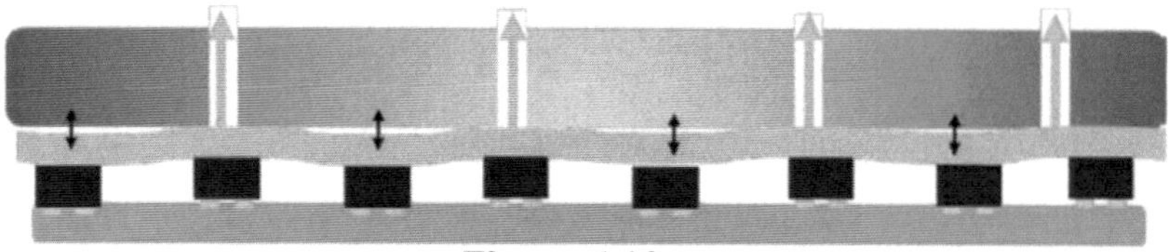

Figure. 4 Airspace

One of the most important issue of wafer direct dipping is how to keep flatness of wafer. It is necessary to vacuum a wafer by area not at pinpoint. We estimated needful additional number of drilling hole is around 80,000. It is quite difficult to meet to such huge number of holes in technically as well as cost wise. So we looked for replacement technology of drilling holes and found out *porous* metal for pusher head. New *porous* head have 141,000 micro vacuum holes which enlarge vacuum area size from 12[mm2] to 1,628[mm2]. See Table 1. comparison of drilled and *porous*. Total vacuuming power is -10.9kPa which is same between drilling and porous due to use same vacuum pomp.

Table 1. Comparison of drilled and *porous*

	Drilled head	Porous head
Number of hole	380	144000
Mm2 per hole	0.031	0.011
Vacuum area size mm2	12	1628
Proportion	1	136

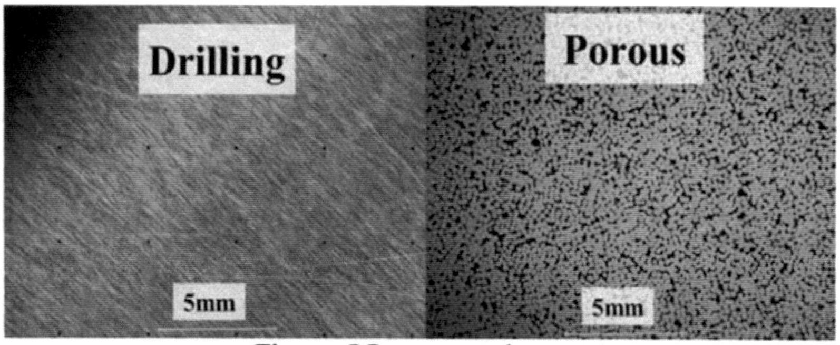

Figure. 5 Porous metal

Porous is made from continuous cracks and gaps, not holes. See Fig. 5 *porous* metal. Thousands vacuum cracks and gaps of *Porous* improved vacuum condition which stabilized vacuum power and brought high level flatness. So we solve existing oversupply or shortage of ACP volume and provide even volume of ACP or other adhesive to thousands mini LED chips. It was removed remaining air between wafer and pusher head by *porous* which led flatness of wafer see Fig. 6 *porous* vacuum and F Fig. 7 dipping marks on ACP pool and evenly ACP dipped to mini LED chips on wafer. We could confirm very highly versatility of wafer direct dipping.

Figure. 6 Porous vacuum

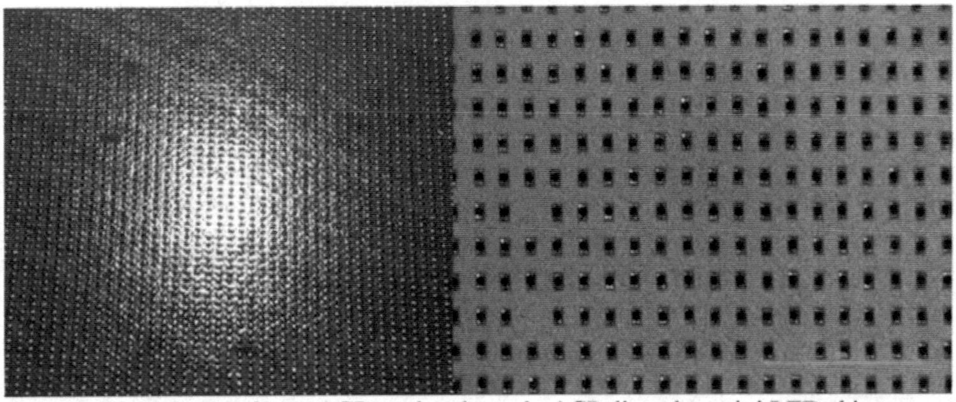

Figure. 7 Dipping marks on ACP pool and evenly ACP dipped to mini LED chips on wafer

3. Introduction [2] Wafer direct chip shooting.

We would like to introduce you method of movement of shooting head of wafer direct dies attachment. Common die bonder pick chip up by vacuum nozzle, then move to substrate and place it there.

Wafer direct chip shooting does not have any pick up motion and no travel to substrate. Ejector pin and plunger are on the same axis but ejector pin is controlled by motor and plunger is controlled by solenoid. We make a pinhole on wafer by ejector pin. Pinhole leads air flow which reduces stickiness of wafer. An adhesive dipped mini LED chip sticks to substrate and transferred it from wafer by return tension of wafer. See Fig. 7 of head unit

1. Ejector pin comes down to top surface of wafer
2. Ejector pin and plunger pushes wafer down and makes a pinhole
3. Transfers mini LED chip to substrate
4. Then ejector pin goes up to home position

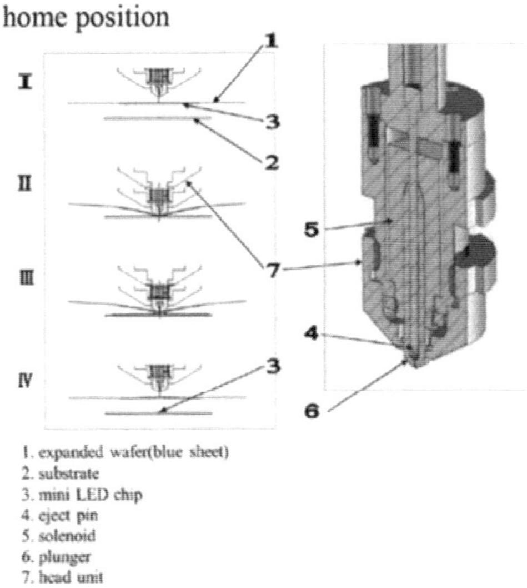

Figure. 8 Head unit

4. Consideration and result [2] Wafer direct chip shooting

We had two technical issues with die attachment process of wafer direct that one is contamination of a broken piece of wafer and other is shift of chip address on wafer due to loosen tension of wafer. Both of them are made by making pinhole. It is appeared a broken piece of wafer on surface of mini LED chip when mount mini LED chip to substrate by making pinhole. This broken piece disturbs lighting of mini LED which must be removed at cleaning process later. Other is shift of chip position. We recognize X, Y axis position of all chips on wafer by camera before starting chip mounting. We use this X, Y coordinates as target position each chip for chip bonding. But chip location is shifted one by one after making pinhole. Because pinhole loosens tension of wafer which shifts original X, Y coordinates of chip. If we recognize X, Y position of chip again during mounting which drastically disturb the productivity. We considered to find the algorithm of loosen tension of wafer by pinhole for prediction of position change. But we could find after test of thousands chip mounting that loosen of tension of wafer is in random and no algorithm is available. We understood that we can't control shift of chip position by tension change of wafer and have to find the condition of pinhole free chip mounting. If we push down mini LED chip by ejector pin only, ejector pin penetrates wafer. We do not get any pinhole when we push down mini led chip by ejector pin and plunger together, but it is not possible to transfer mini LED chip from wafer to substrate due to that adhesion of wafer is stronger than adhesion of ACP or solder. Ejector pin and plunger together to push down mini LED and keep it for 5milliseconds. Wafer leaves from mini LED by return tension except holding area by ejector pin when plunger goes up. Ejector pin still stays at bottom further 4milliseconds to well transfer mini LED to substrate by surface tension of adhesive without making any pinhole. See Fig. 9 structure of pinhole free chip shooting.

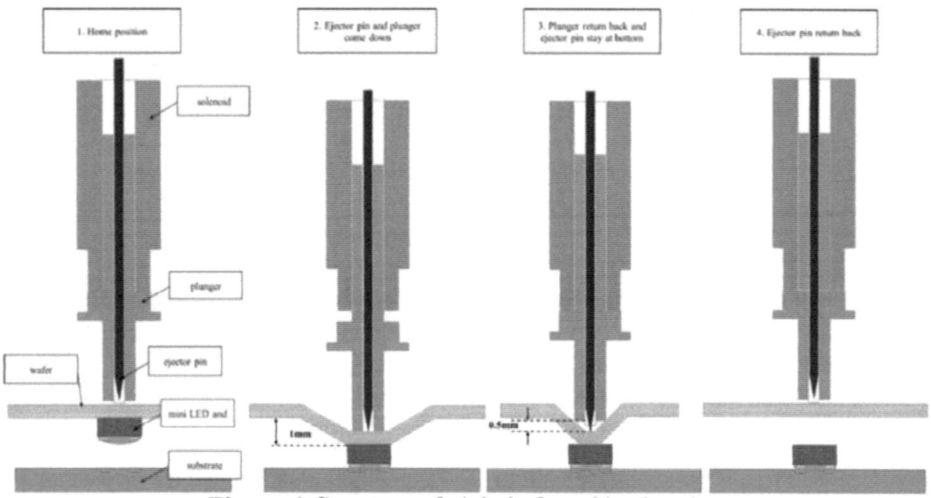

Figure. 9 Structure of pinhole free chip shooting

Ejector pin, plunger and solenoid are assembled on Z axis which is under control by Z axis motor. Plunger is assembled on Z axis but is under controlled by solenoid. Key technology of pinhole free mini LED chip mounting is movement of plunger and waiting time. See below Fig. 10 timing chart of plunger and ejector pin.

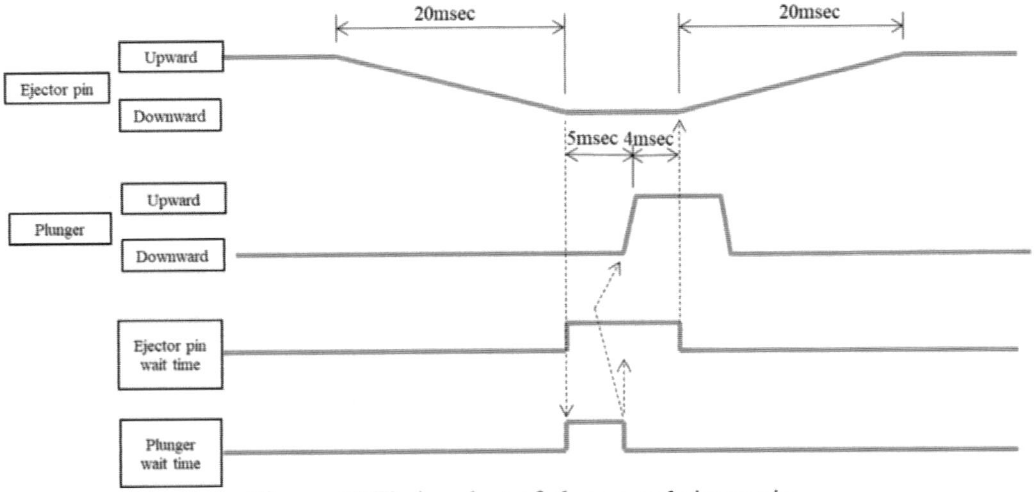

Figure. 10 Timing chart of plunger and ejector pin

We could confirm that both of ACP and solder paste well separated and aggregated under wafer direct dipping technology. We had no solder bridge and no remaining solder ball at gap. See Fig.11 X-ray after reflow. ACP is mixture of solder and epoxy resin which has "self-alignment" function. Melted solder in ACP flows to P electrode and N electrode of chip and substrate in even volume under reflow. This flow power is strong as called as "self-alignment" which corrects mounting offset of mini LED chip. Solder paste in screen printing has same function but it has weak offset correction due to that screen printing supplies minimum volume of solder to electrodes of substrate which does not flow. But wafer direct dipping supplies ACP in the center of mini LED chip that solder flow from center to edge in long distance which enlarges self-alignment" function. See Fig. 12 wafer direct dipping result. So we can see good *"self-alignment"* function with solder paste to be supplied by wafer direct dipping. This is advantage of wafer dipping to screen printing.

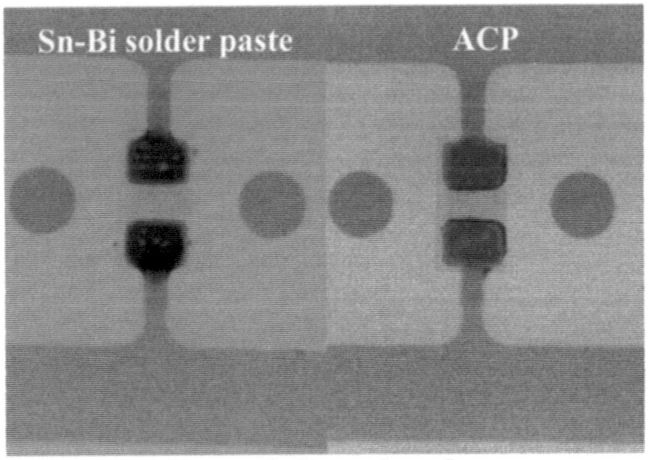

Figure. 11 X-ray after reflow

Figure. 12 Wafer direct dipping result

It is not suitable to use ACP in screen printing due to that epoxy resin of ACP is sticky which is not removable from the stencil. But it is possible to use both of ACP and solder paste in wafer direct dipping. See Fig. 12 wafer direct dipping result. So we can say that wafer direct dipping technology can optimize supply volume of any adhesive. Wafer direct dipping can replace existing screen printing of solder paste on mini LED die bonding.

Table 2. Comparison chart

	Die bonder	Wafer direct
Pinhole	Yes	No
Pin damage	Yes	No
Contamination	Yes	No
Touch electrode	Yes	No
Flip chip	No	Yes

Common die bonder for mini LED bonding is still having not escapable several potential problems such as damage to electrodes of mini LED chip by ejector pin and remaining debris of wafer on electrodes. Wafer direct chip shooting with pinhole free technology can solve existing such two problems of die bonder in the same time. See Table. 3 comparison chart. Wafer direct chip shooting can achieve the speed of 0.098seconds per mini LED chip, and X, Y axis accuracy is ±0.01mm after reflow. See Table. 3 bonding accuracy.

Table. 3 Bonding accuracy

	Sn-Bi solder		ACP	
After reflow	X	Y	X	Y
Ave.(mm)	-0.004	0.000	0.003	0.000
3 sigma(mm)	0.013	0.016	0.012	0.011

You can see Figure. 13 [10,050] chips placed on lighting module. We do not have any unlit chip on this module. We can say that yield rate of wafer direct technology is high that our mini LED chip mounting rate is 99.7% or more. We would like to introduce details of our wafer direct technology for improvement of die attachment technology of mini LED flip chip.

Figure. 13 Lit 10,050 chips on lighting module

Acknowledgments

We got technical suggestion and advice from below two gentlemen for development of wafer direct technology.

Dr. J.T. Moon of Hojeonable Co., Ltd. Korea for ACP and adhesive technology

Dr. Beng So Ryu of QMC Co., Ltd. Korea for pinhole free placement technology

5.References

[1] Patent No. WO-A1-2012/139766 dated 20121018 "Device and method for transferring electronic components from a first carrier to a second carrier" owned by MÜHLBAUER AG, Denmark.

[2]Paten No. US9633883B2 dated 20170425 "Apparatus for transfer of semiconductor devices" owned by Rohinni LLC, USA.

[3] Patent No. JP6603412 "Mounting method, mounting head and mounting apparatus" owned by T. Tatsuiwa, Japan.

[4]Patent No. JP6543421 "Transfer method and mounting method" owned by T. Katashio, Japan.

[5] P205-202007221429 "Optimization and quantification of "mini LED flip chip on board (COB) package" by self-alignment with anisotropic conductive paste (ACP) and pinhole free wafer direct technology" on November 25, 2020 for " PROCEEDING(SSLCHINA & IFWS 2020). Published by China solid state lighting alliance and China advanced semiconductor industry innovation alliance. Page No. 194 by K. Hiraki.

Journal of Physics: Conference Series **2065** (2021) 012016 doi:10.1088/1742-6596/2065/1/012016

A Flexible Solar-blind Ultraviolet Photodetector based on Carbon Nanodots

Mengru Zhu, Zhiwei Zhao*, Yong Fang, Zhengjin Weng, Wei Lei

School of Electronic Science and Engineering, Southeast University, Nanjing 210096, P. R. China

E-mail: Zhao_zw@seu.edu.cn

Abstract. A flexible solar-blind ultraviolet (UV) photodetector based on the carbon nanodots (CNDs) was fabricated on a polyethylene terephthalate (PET) substrate. The responsivity of 1.2 mA/W is obtained for 10 V applied bias under 254 nm illumination. Further, bending tests were carried out under the 0.2% strain, and the results showed that this flexible photodetector had stable characteristics and no obviously decrease of the photocurrent. The bending performances exhibit excellent potential for the fabrication of smart and flexible photodetectors.

1. Introduction

Due to the strong absorption by ozone layer, UV radiation with wavelength between 200nm to 280nm has almost no background interference on the earth surface, so it also called solar-blind UV radiation. Based on this characteristic, solar-blind UV photodetectors have attracted more and more researchers with the strong anti-background radiation interference ability and high detection sensitivity [1,2]. Conventional solar-blind UV photodetectors are fabricated on bulk silicon, which hinder their application on the flexible electronic device field. Compared with the photodetectors based on rigid substrates, flexible photodetectors have wide application in biomedical, portable and wearable devices. Sensitive photo-response and high mechanical strength are essential for a high-performance flexible photodetector. From the perspective of detection capability, high detection ability in the target wavelength is very important for the active layer material. In terms of mechanical strength, the flexible device should maintain high sensitivity under repeated bending conditions. In general, in order to meet the requirement that the performance of the flexible photodetector will not be significantly reduced after repeated deformation, the active material also needs to be able to withstand repeated deformation. According to the tensile bending strain equation, the strain of the material is proportional to the thickness of the material [3]. Under a given radius of curvature, the material with thinner thickness has the smaller strain. Therefore, low-dimensional nanomaterial is a perspective material to fabricate the high-performance flexible photodetector. As a kind of low-dimensional nanomaterial, zero-dimensional quantum dots can be synthesized by simple solution method [4, 5]. Compared with photodetectors based on traditional bulk materials, photodetectors based on quantum dots generally have higher response and gain due to the high specific surface area and small size. Therefore, using quantum dots as the active layer material can not only strike a balance between the photo-response and flexibility, but also improve the performance of the device and simplify the fabrication process.

Carbon nanodots (CNDs) are quasi-spherical nanoparticles with the diameter below 10 nm, which have excellent properties including outstanding chemical inertness, low toxicity and fascinating optical properties [6]. Based on quantum confinement and edge effects, the band gap of CNDs is tunable [7].

Studies have found that CNDs prepared by electrochemical methods have strong absorption at a wavelength of about 230 nm [8], but have almost no absorption in the visible and infrared regions. This means that CNDs are very suitable for preparing solar-blind photodetector. Besides, the solvent of the CNDs is aqueous solution and it can be transferred to the substrate by spin coating. Since most of the common flexible substrates are not resistant to high temperature and organic solvent, it is a good choice to use CNDs as the active layer of flexible photodetectors.
Therefore, this work report on a high-performance solar-blind flexible photodetector based on CNDs as an active layer and polyethylene terephthalate (PET) as a flexible substrate. This work also provides a new idea for the use of low-dimensional nanomaterials to construct high-performance, low-cost, and environmental-friendly flexible photodetectors.

2. Experimental

2.1 CNDs synthesis
The CNDs were prepared by electrochemical method at 60V bias voltage. Two carbon rods (99.99%, length of 10 cm, diameter of 0.6 cm) were vertically fixed at a distance of 2cm as electrodes to react in distilled water. The reaction was carried out with continuously magnetic stirring. Finally, the obtained black-brown liquid was filtered to remove large particles, and then centrifuged at 10000 rpm for 30 minutes. The suspension is as-prepared CNDs solution.

2.2 Device fabrication
The Schematic diagram of the photodetector is shown in Figure 1. The PET (thickness of 0.175mm) was ultrasonically cleaned for 30 minutes and UV ozone treated for 30 minutes. Aluminum (Al) electrodes were prepared using thermal evaporation on PET substrate. The as-prepared CNDs solution was mixed with PVA solution (3% in weight in water), and then spin-coated (500 rpm for 10 s and 2000 rpm for 30 s) between the electrodes as the active layer, and the channel width was 50 μm. Finally, the device was annealed at 50 °C for 10 minutes.

2.3 Characterization
Transmission electron microscopy (JEOL-02100, Japan) was used for morphology of the CNDs. The UV-visible absorption spectrum was measured by UV-visible spectrophotometer (UV-2600, Shimadzu). Photo-response measurements were performed using a mercury lamp (254nm, OSRAM G11T5/OF) and a semiconductor characterization system (Keithley 4200-SCS).

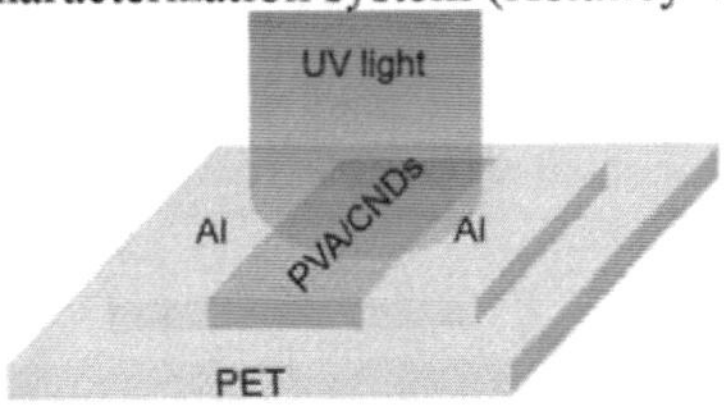

Figure 1. Schematic diagram of the flexible PVA/CNDs photodetector.

3. Results and discussion
The morphology of the as-prepared CNDs is shown in Figure 2a. It can be seen that the CNDs are quasi-spherical and have an average diameter of about 3.3 nm. It is observed that the absorption of CNDs mainly in the ultraviolet region, and gradually weakens with the increase of wavelength (Figure 2b). Moreover, a main peak at about 230 nm can be observed which corresponds to the π-π* transition of C=C bonds of CNDs [9]. PVA/CNDs has the same absorption peak as CNDs, indicating that introduce PVA into CNDs will not affect the absorption performance. The optical spectra was further analyzed to calculate the energy band gap. The absorption of CNDs follows a power law of the following form [10]:

$$(\alpha h\nu)^2 = B(h\nu - E_g) \tag{1}$$

where α is the absorption coefficient, $h\nu$ is the energy of the incident photon, B is a constant, and E_g is the energy band gap. Through the inset of Figure 2b, the band gap of CNDs is about 4.6 eV.

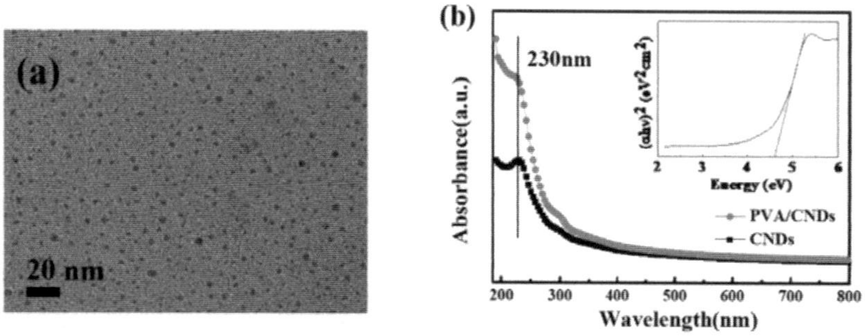

Figure 2. (a) TEM image of CNDs; (b)UV-visble absorbance spectrum of CNDs and PVA/CNDs with the plot of $(\alpha hv)^2$ versus hv for CNDs in inset.

Figure 3a shows the voltage dependent dark and photocurrent (254 nm) of the photodetector. The power density of 254 nm UV light was kept at 1.5 mW/cm^2. It can be clearly seen that under dark and illumination, the currents increase linearly with the increase of the applied bias voltage. Such linear relationship indicates that there is a good ohmic contact between Al and PVA/CNDs film. Further, the photocurrent significantly increased compared with the dark current. In order to study the response time of the photodetector, time-dependent photocurrent measurement was also performed at 10 V bias voltage. As shown in Figure 3b, the flexible photodetector has stable switching response. The current increases when the light is irradiated, and recovers after the light is removed. It is worth noting that the dark current of the device is about 1nA, which is lower than many flexible solar-blind photodetectors [11].

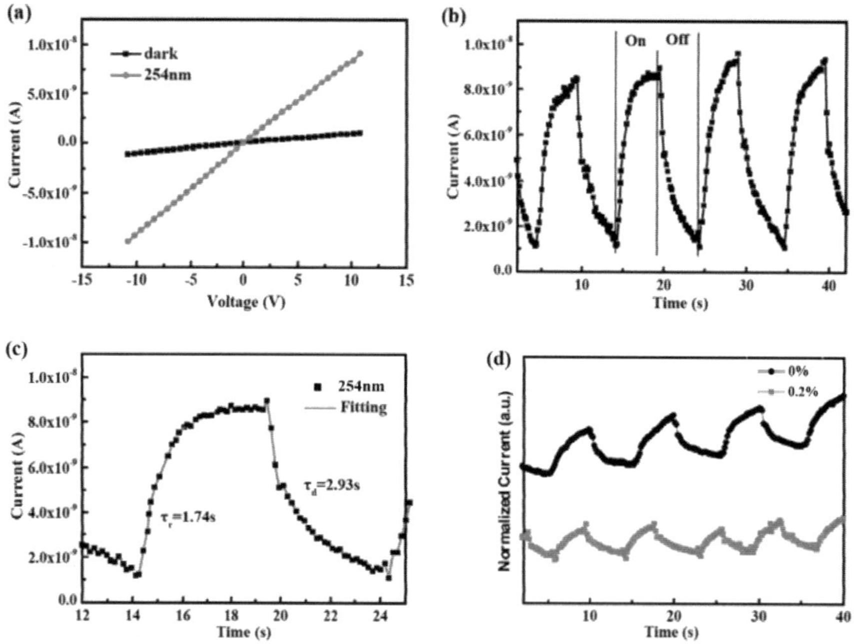

Figure 3. (a)Variation of dark current and photocurrent with applied bias to the photodetector; (b)time response curve of the photodetector; (c) experimental curve and fitting curve of the current rise and decay process; (d) time response curves under different strain.

For photodetectors, response time is defined as the time interval between 10% and 90% of the response peak. Therefore, the rise time of the device is about 1.74s and the decay time is 2.93s (Figure 3c). The decay time is little slower than the rise time, which can be attributed to the traps in the active layer. When illuminated by UV light, electron-hole pairs are generated in CNDs. Some of the carriers are captured by the trapping states in PVA/CNDs films. After turning off the light, the trapped carriers are released, which contributes to the slow-response. In addition, responsivity (R_λ) is an important parameter to measure the response performance of photodetectors, which can be calculated by the following formula [11]:

$$R_\lambda = \Delta I/PS \tag{2}$$

where ΔI is the photocurrent of the device, P is the power of the light source, and S is the effective area of device. The S of 5×10^{-3}cm^2 was considered for the calculations. The responsivity of about 1.18mA/W was acquired at 10 V bias.

In order to study the performance of the device under bending, the photoresponse of the device under different bending strains was tested. The strain (ε) in the device was calculated by the following formula [12]:

$$\varepsilon = (t/2R) \times 100\% \tag{3}$$

Where t is the thickness of the device and R is the radius of curvature. The switching characteristics of the device under strains of 0% and 0.2% were tested, as shown in Figure 3d. It can be clearly seen that the photocurrent of the device under 0.2% strain is almost the same as the photocurrent of 0% strain, and the device also has obvious photoresponse and stable switching characteristics. This indicates the flexible photodetector based on CNDs can successfully detect solar-blind UV light under a bending strain.

4. Conclusion

In conclusion, a flexible solar-blind UV photodetector was successfully prepared based on high-quality CNDs prepared by an electrochemical method. The as-prepared CNDs have good absorption in the solar-blind UV range. Under the 254 nm illumination, the device shows a responsivity of 1.18 mA/W at 10 bias voltage. Under the bending strain, the performance of the device did not deteriorate significantly, demonstrating the potential applicability in fields that cannot be served by conventional rigid photodetectors.

Acknowledgments

This work was financially supported by the National Natural Science Foundation of China (51879042, 51839011); Postgraduate Research & Practice Innovation Program of Jiangsu Province, the Fundamental Research Funds for the Central Universities (KYCX20_0105); State Key Laboratory of Nuclear Power Safety Monitoring Technology and Equipment (K-A2020.415); National Key Research and Development Program of China (2018YFE0125500).

5. Reference

[1] Li L, Auer E, Liao M, et al. 2011 *Nanoscale.* **3**(3):1120-1126
[2] Mitra S, Pak Y, Xin B, et al. 2019 *ACS Appl. Mater. Interfaces.* **11**(42)
[3] Chen H 2019 *Appl. Phys.* **09**(8):373-378
[4] Yang J X, Wang J, Lim A, et al. 2009 *Acs Nano.* **3**(8):2367-75
[5] Pan D, Xi C, Li Z, et al. 2013 *J. Mater. Chem. A* **1**(11):3551-3555
[6] Goncalves H, Jorge P A S, Fernandes J R A 2010 *Sensors Actuat. B Chem.* **145**(2):702-707
[7] Sun Y P, Zhou B, Lin Y, et al. 2006 *J. Am. Chem. Soc.* **128**(24):7756-7757
[8] Li X, Rui M, Song J, et al. 2015 *Adv. Funct. Mater.* **25**(31):4929-4947
[9] Zhang J, Abbasi F, Claverie J. 2015 *Chem.* **21**(43): 15142-15147
[10] Kumar S S, Rubio E J, Noor-A-Alam M, et al. 2013 J. Phys. Chem. C **117**(8):4194-4200
[11] Tak B R, Gupta V, Kapoor A K, et al. 2019 *ACS Appl. Electron. Mater.* **1**，2463-2470
[12] Han K L, Jeong H J, Kim B S, et al. 2019 *J. Phys. D: Appl. Phys.* **52**(48):483002.

Research on Thermal Fatigue Life Prediction Technology of PoP Laminated BGA Products

Hong-lei Ran*, Xiao-jie Sheng, Qiang Duan, Kui Zhang, Jie Huang

National Semiconductor Device Quality Supervision and Inspection Center, The 13th Research Institute of CETC, Shijiazhuang, China

hongleiran@163.com

Abstract. In this paper, a fast prediction method for thermal fatigue life of PoP laminated BGA Product was proposed. Firstly, the stress and strain of the solder joints of each layer of the laminated device in thermal fatigue test was determined by finite element simulation method. According to the research idea of relative stress and strain, the solder joints were divided into sensitive solder joints and reliable solder joints. Secondly, sensitive solder joint were connected with PCB traces through the internal pads, bonding wires, TSVs and reliable solder joints to form a daisy chain. Through real-time dynamic monitoring of the resistance change of the daisy chain in thermal fatigue test to judge whether the solder joints fail, and record the occurrence time of the first failure solder joint. Finally, the thermal fatigue life of the product was estimated by the Norris-Landzberg formula.

1. Introduction

With the light, thin, small and functional diversification of electronic products, ball grid array package (BGA) has become the most advanced packaging technology in integrated circuits[1-3]. Products based on BGA packaging technology have been fully penetrated into all sectors of the national economy. The failure of BGA packaging products will cause huge economic losses in various field of our country. According to statistics, more than 50% of the failures of BGA products are caused by welding assembly failure, and field failure studies show that 60% of the failure mechanisms can be attributed to thermal fatigue damage. How to calculate and analyze the residual stress in different solder joints of BGA products, so as to accurately predict the service life of solder joints, is one of the current research hotspots[1-3].

Package on package (PoP) is an innovative result of packaging industry. The vertical integration of two independent BGA products can significantly increase the system packaging density. It has many advantages such as convenient combination, flexible design and low signal delay[1-3]. Compared with single-layer BGA products, the research on thermal fatigue life of PoP laminated BGA products is almost blank. Although the process nature of PoP packaging products still belongs to surface mount technology (SMT), due to the more complex effects of temperature residual stress, warpage deformation and greater probability of solder ball defects, the relevant research results of single-layer BGA products cannot solve the technical problem of service fatigue life of PoP laminated BGA products[4].

In order to solve the technical problem of thermal fatigue life of PoP laminated BGA products, a 3D structure of solder joint reliability monitoring daisy chain structure is proposed, and a real-time online monitoring system for solder joint thermal fatigue life is designed. By dynamically monitoring the

Journal of Physics: Conference Series **2065** (2021) 012017 doi:10.1088/1742-6596/2065/1/012017

resistance of each link of daisy chain, the failure solder joints can be quickly found, and the failure time can be recorded timely, which is simple and efficient[5].

2. Analysis of research methods for thermal fatigue life

The research on the reliability of SMT products usually focuses on two aspects, namely, the reliability of electronic components and the reliability of solder joints. As the solder joints of SMT product are the weakest link in the entire product, the failure of any solder joint may lead to the failure of the product. Therefore, the research on the reliability of products such as BGA products mainly focuses on the reliability of solder joints, especially the thermal fatigue life of solder joints.

According to different research objects, the research methods for thermal fatigue life of solder joints of BGA products can be divided into three categories[4][5][8]. The first one is to study the packing system, which includes devices, solder joints and mounting substrate. This kind of analysis method usually uses simulation method to apply temperature cycling load on the packaging system to simulate its service state. Because it is practical to model a large number of solder joints with different sizes, the system modeling is usually simplified, or the mesh is simplified. In addition, the simulation analysis method cannot fully consider the influence of product assembly process, residual stress, device size and the location distribution of solder joints in the assembly process will affect fatigue life of the solder joint, so such modeling analysis cannot reflect the real situation of solder joint fatigue. The second category is to conduct research on the device, that is, to carry out temperature cycling test on device and auxiliary other means such as solder ball shear test for the device containing solder ball. This kind of method also does not consider the influence of product assembly process. The third type is to conduct research on the pure solder joint, which needs to simulate the solder joint sample, so the analysis results are not representative. Therefore, the current research methods on the thermal fatigue life of BGA products are more or less defective, and the related research on the thermal fatigue life of the solder joint of PoP laminated BGA products is even less. To sum up, the optimal method to study the thermal fatigue life of BGA solder joints is to apply physical thermal stress cyclic load to the research object of packaging system level.

The results show that the thermal fatigue damage of solder joints is mainly caused by the cyclic plastic strain of solder joints[6-8]. Based on the plastic strain theory, Coffin and Manson proposed and improved the prediction method based on the experimental data, and established the Coffin-Manson relationship between the load cycle and the plastic strain amplitude when fatigue failure occurs. However, the influence of time and temperature on life of the solder joint is ignored in this model. In order to overcome the above shortcomings, the EngelMaier-Wild model and the Norris-Landzberg model improved based on the Coffin-Manson formula are currently widely used. The EngelMaier-Wild model mainly considers the relative displacement between BGA product and the mounting substrate due to different coefficient of thermal expansion (CTE), which is suitable for the case where the shear force is the principle stress of solder joints. The model is mainly based on the finite element analysis, assuming that the assembly process is in an ideal state, without considering the effect of assembly process on the thermal fatigue life of solder joints. Therefore, this method can be used to quickly evaluate the thermal fatigue life of solder joints in the early stage of design, and judge whether the designed products meet the system-level working requirements. The Norris-Landzberg model is mainly based on the accelerated test results, combined with the service conditions to calculate the thermal fatigue life of solder joints, so this model is in line with the research idea of "apply physical thermal stress cyclic load to the research object of packaging system level".

The Norris-Landzberg formula is a thermal fatigue life model based on physical temperature cycling test. Assuming that the plastic strain range is linearly proportional to the temperature range, the actual thermal fatigue life is calculated by using the maximum temperature, cycle frequency and other parameters of accelerated test and practical application conditions. The details are as follow.

$$N_f = N_t \times A.F \tag{1}$$

N_f is the predicted actual thermal fatigue life of solder joints, N_t is the number of cycles of the accelerated test, and $A.F$ is the acceleration factor.

$$A.F = \left[\frac{f_f}{f_t}\right]^m \times \left[\frac{\Delta T_f}{\Delta T_t}\right]^{-n} \times \exp\left[\frac{H}{K} \times \left(\frac{1}{T_{\max f}} - \frac{1}{T_{\max t}}\right)\right] \quad (2)$$

H is the activation energy of solder, ΔT_f is the application temperature range of product service process, ΔT_t is the test temperature range, K is Boltzmann constant, f_f is the temperature cycling frequency of product service process, f_t is the cycle frequency of physical simulation test. $T_{\max f}$ is the highest temperature (unit: K) during service process, and $T_{\max t}$ is the highest temperature of physical simulation test.

3. Design of Daisy Chain for Thermal Fatigue Life Monitoring of PoP Laminated BGA Products

3.1 Analysis of general monitoring methods

The thermal fatigue failure of BGA solder joints usually results in cracks or micro cracks, which are characterized by electrical connection open circuit or increased resistance[8][9]. In general, the life monitoring method of thermal fatigue life is to carry out temperature cycle test on the product to be examined, take out the product after reaching the predetermined cycle number, and determine whether the product is abnormal through auxiliary means such as electrical test. The result of electrical performance test determines whether to carry out further temperature cycling test or failure analysis used to determine whether it is solder joint failure. This kind of method can be called "interruption monitoring method". If the electrical characteristics of the product are abnormal, non-destructive inspections, such as visual inspection and X-ray inspection, are usually carried out first. Visual inspection can only defect the solder joints at the edge of the device, but not the solder joints hidden at the bottom of the device and the defects inside the solder joints. X-ray inspection uses X-ray projection characteristics to further detect the failure solder joints hidden at the bottom of the device, but it is likely that the defects with slight deformation but also affect the reliability of the product cannot be inspected. Further detections are destructive detection techniques, including dye penetration, sample preparation grinding, scanning electron microscope (SEM) observation and other analysis means, which can usually identify most of the solder failure joints, but destructive detection generally requires a large number of sample preparations, and the analysis process is expensive and time-consuming. As an organic whole, the thermal fatigue life of BGA products is marked by the failure of the first solder joint. Due to the limitations of the research methods, the "interruption monitoring method" usually cannot accurately located the first failed solder joint. In addition, the micro cracks of some failed solder joints increase under high temperature stress, and recover after unloading high temperature stress. Because the "interruption monitoring method" can only carry out static analysis of products, it cannot solve this problem.

Daisy chain is another important means to monitor the thermal fatigue life of BGA products. The internal pads of BGA products are connected by bonding wires, and then the external pads are connected by the printed circuit board (PCB) to form a network structure. The reliability of package welding is studied by monitoring the change of resistance value of the daisy chain. Because the solder joint is the weak link of the entire daisy chain, when the resistance value changes beyond a certain range, it can prove that the solder joint has quality problems. Daisy chain method can overcome the disadvantages of the "interrupt monitoring method". However, at present, the daisy chain method only aims at the products with singe-layer packaging structure, so this method cannot be directly applied to the PoP laminated BGA products in this paper.

3.2 Simulation and analysis of sensitive solder joints of PoP laminated BGA products

Due to the difference of CTE between the BGA product and the mounting substrate, the thermal deformation of the two is different when the temperature changes, resulting in stress. The figure below is the schematic diagram of a certain type of PoP laminated BGA product after assembly.

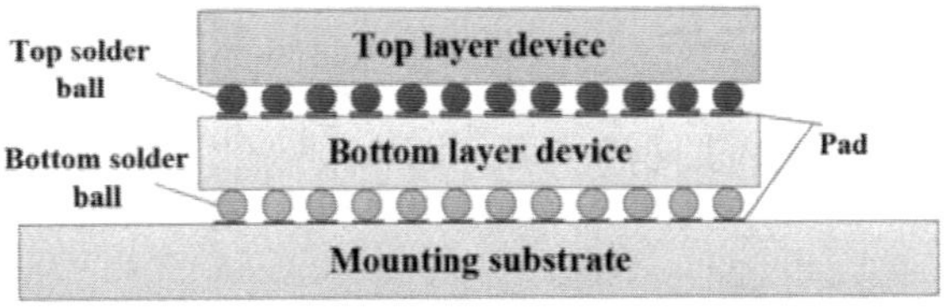

Figure 1. The diagram of assembly product.

The vertical signal interconnection is realized by through silicon via (TSV). The TSV technology is relatively mature, and its reliability is much higher than that of BGA solder joint under the same stress[10-11]. The top device is welded and assembled with the bottom device through the top solder ball, and the assembled product is assembled to the mounting substrate through the solder ball of the bottom device. The material is shown in the table below.

Table 1. Table Of Material Parameters

Structure name	***material***	**Structural properties** *Young modulus (Gpa)*	*Poisson Ratio*	*CTE (ppm)*
device	Si	129	0.21	3.5
Solder ball	63Sn37 Pb	43	0.36	25.4
pad	Cu	58	0.30	17.7
Mounting substrate	FR4	7.0	0.23	15.8

Through the finite element simulation method, the stress of the product shown in Fig.1 was analyzed under the condition of temperature cycling. The elastic-plastic model is used for the solder balls and pads, and the shell model is used for the mounting substrate. The results with the thermal load condition -55°C is shown in Fig.2, where Fig.2 (a) is the product stress distribution diagram, and Fig.2 (b) is the diagram of stress distribution of the bottom solder balls.

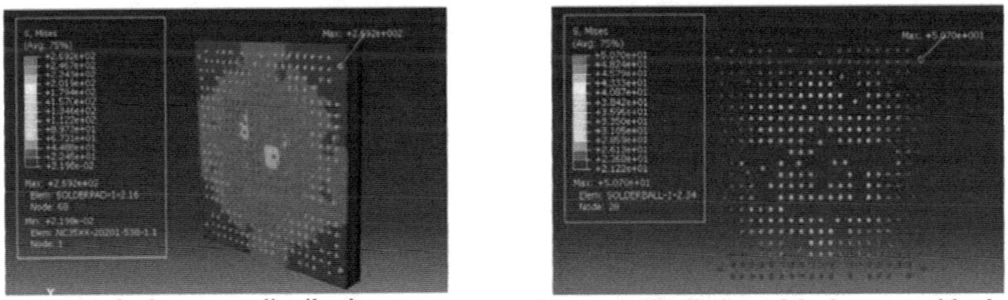

(a) device stress distribution (b) stress distribuion of the bottom solder ball

Figure 2. Stress distribution diagram.

Due to the different CTE of materials, the deformation of the assembled products will produce stress shown in Fig.2 during the temperature change process, and the edge stress of both the top layer device and the bottom layer device is larger than that of the middle region. Because the top layer device and the bottom layer device are made of silicon, the relative deformation between the top layer device and the bottom layer device due to CTE difference is less than the relative deformation between the bottom layer device and the mounting substrate, so this paper takes the bottom layer device as an example to analyze. The device is a 20mm square. For the solder ball distribution surface of the bottom layer device, the coordinate system is established with its geometric center as the origin. For the interface with y=0mm, the result of stress-strain analysis is shown in the figure below.

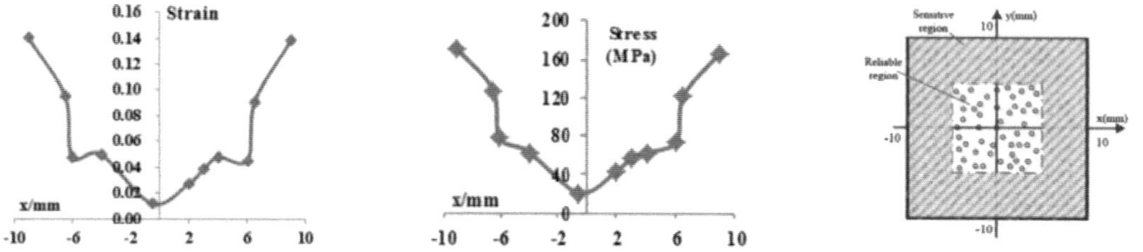

(a) Strain distribution (b) Stress distribution (c) Sensitive region

Figure 3. Division of sensitive solder joint and reliable solder joint.

Because the thermal fatigue life of BGA products is marked by the failure of the first solder joint, this paper adopts the research idea of relative stress, and the failure probability of high stress solder joint is greater than that of low stress solder joint. Assuming that the highest stress in the solder joints is σ, the solder joints with stress between σ and $\sigma/N(N\geqslant 2)$ are considered as suspicious solder joints, that is, sensitive solder joints, and the solder joints with stress less than σ/N are considered as reliable solder joints. If the sensitive solder joints are reliable during the temperature cycling test, the whole product is considered to be reliable in the application environment. In this paper, N=2 is chosen. Meanwhile, in order to facilitate the research and improve the reliability margin, the sensitive region is expanded and simplified to a rectangular ring as shown in Fig.3(c), whose length is 1/4 of the corresponding side length. Therefore, the research on thermal fatigue life of PoP laminated BGA products focuses on the sensitive solder joints in the rectangular ring.

3.3 Daisy chain design of PoP laminated BGA product

As the failure of any solder joint will lead to the failure of the product, the thermal fatigue life of PoP laminated BGA products must be evaluated for all solder joints.

According to the research content of *Chapter B*, this paper only need to consider sensitive solder joints, which can significantly reduce the workload, save monitoring resources and improve work efficiency. Because the edge solder joints of two stacked devices are sensitive solder joints and the middle region is reliable solder joints, the sensitive solder joints of the top layer device can be connected with the PCB lines through the internal pads of both the top layer device and the bottom layer device, bonding wires and reliable solder joints of both top solder joints and bottom solder joints to form a daisy chain network. Since the PoP laminated BGA product is connected with the PCB lines through the bottom solder balls, the final lead-out ends of the daisy chain are reliable solder joints in the middle region of the bottom device. This type of daisy chain is defined as the first type of daisy chain in this paper. Compared with the sensitive solder joints of the top layer device, the daisy chain of the bottom layer device can be completed only through the internal pads of the bottom layer device, bonding wires and sensitive solder joints of the bottom layer device, which is defined as the second type of daisy chain in this paper. The schematic diagram of the first type of daisy chain and the second type of daisy chain is shown in Fig.4.

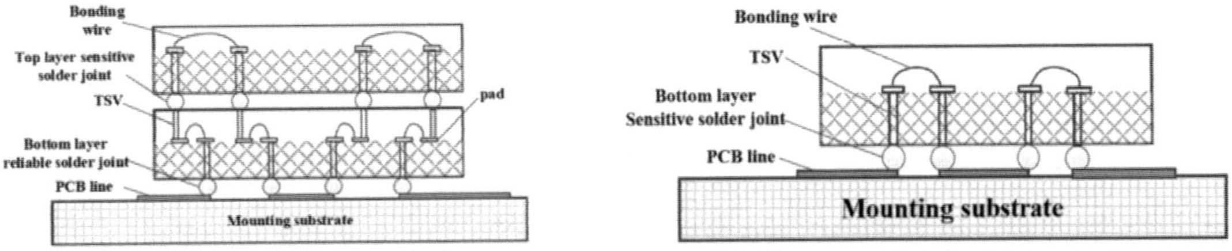

(a) The first kind of daisy chain (b) The second kind of daisy chain

Figure 4. Schematic diagram of daisy chain.

3.4 Design of on-line monitoring system

In order to accurately monitor the thermal fatigue life of PoP laminated BGA products, an on-line monitoring system for the thermal fatigue life of solder joints is designed as shown in Fig.5, in which the high-temperature connector and high-temperature cable can withstand the high temperature above 150℃. The scheme can also be extended to other environmental reliability test, such as vibration reliability.

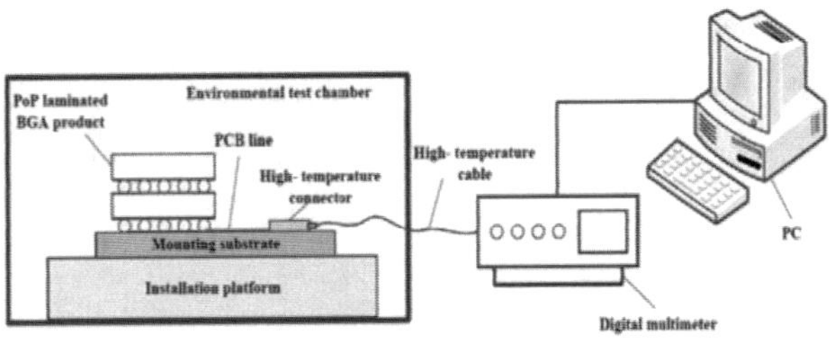

Figure 5. On-line monitoring system.

Figure 6 is the schematic diagram of daisy chain resistance on-line monitoring. Through dynamic monitoring the resistance change of the daisy chain in real time, the solder joints that failed in the thermal fatigue test can be quickly found. The change of resistance value of R1, R2 and Rn in local daisy chain can quickly locate the failure solder joint, provide corresponding information for the possible failure analysis test and assist product design improvement. Designers can carry out overall or partial combination daisy chain resistance monitoring such as R in Fig.6 as required, and achieve the purpose of assessing the reliability of regional solder joints. The reasonable use of the partial combination daisy chain resistance can also reduce the monitoring resources and the cost.

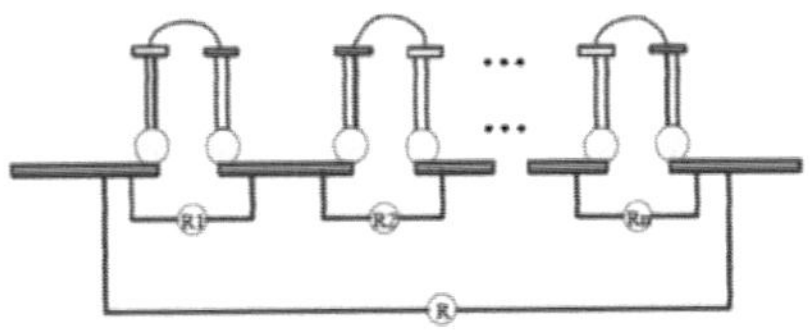

Figure 6. On-line monitoring system.

The position of the failed solder joint and the time of the first failure solder joint can be quickly located by the on-line monitoring system of thermal fatigue life. Then, according to the formula Norris-Landzberg and the actual service environment of the product, the thermal fatigue life can be calculated quickly.

4. Conclusion

This article extends the traditional daisy chain research method of solder joints reliability of single-layer BGA products to PoP laminated BGA package products. Through finite element simulation, the solder joints of PoP laminated BGA products are divided into sensitive solder joints and reliable solder joints, and the reliability of the products is studied by establishing a daisy chain network. Although this paper focuses on double-layer BGA devices, this method can also be applied to three or more layers of BGA package or other interconnected stack devices. Practice shows that this method is simple and efficient, and can quickly obtain the thermal fatigue life of laminated BGA products.

5. References

[1] Chen Yi-long. "Analysis of the Reliability and Fatigue Life of BGA Solder Joints". Xidian University, 2016.

[2] Shao Chen-xi, "Heat transfer Characteristics of electronic device and its fatigue analysis". Southeast University,2016.

[3] Ji Qing, "Construction analysis of Domestic filp for aerospace". Shanghai Jiao Tong University,2019.

[4] Wei He-lin, "Research on the thermal fatigue reliability and interfacial reactions of plastic ball grid solder joints", Beijing University of Chemical Technology,2011.

[5] Ran Honglei, Wei Ting, Zhang Kui, Huang Jie, Liu Huaguang, Zhao Hailong, et al. "Welding relaibility evaluation method for stacked BGA," Semiconductor Technology Vol.46, in press.

[6] Wang Hui, Guan Yan, Chen Hao, A thermal fatigue life prediction study for CCGA solder joints, Quality and Reliability,Vol.2. pp.31-34.

[7] Reza Ghaffarian, "CCGA Packages for space application, Microelectronics reliability,2006.Vol.46, PP.2006-2024.

[8] Su L, Shi T, Liu Z, et al. "Nondestructive diagnosis of flip chips based on vibration analysis using PCA-RBF", Mesh Syst Signal Process,2017,85,PP.849-856.

[9] Lu X N, Liu F, He Z Z, et al. "Defect inspection of flip package using SAM technology and fuzzy C-means alrorithm", Science China ,2018,61(9),pp.1426-1430.

[10] Ran H L, ZHANG K, HUANG J, et al. "Research on a new test method for through silicon via channel". Proceedings of the 12th international workshop on electromagnetic compatibility of integrated circuits. Hangzhou, China, 2019,pp.87-90.

[11] Nannan Li, Honglei Ran, Chaoyang Xing, Peng Sun, Zhengqiang Zhu, Fanchen Meng,"Fabrication of SOI based inerposer with Air Gapped Si-TSV for 2.5D integration", IEEE 20th international conference on Electronic Packaging Technology, ICEPT 2019, Hongkong, China, 4pps.

Journal of Physics: Conference Series **2065** (2021) 012018 doi:10.1088/1742-6596/2065/1/012018

Influence of Different Surface Passivation on Photoelectric Properties PbS QDs

Yingxiang Yang[1,*], Hongbin Pu[2], Xu Bei[2], Chunlan Chen[2], Min Wang[2], Xiamiao Hu[1]

[1]Design institute, Shaanxi Yellow River Group Co. Ltd, Xi'an,China
[2]Institute of Automation and Information Engineering, Xi'an University of Technology, Xi'an,China

*yangyxx@foxmail.com

Abstract. The photoelectric characteristic of ITO/PbS quantum dots (QDs)/Al structure based single and hybrid iodine ions(I^-) passivated PbS QDs have been investigated. The ITO/PbS QDs/Al structure with hybrid I^- passivation have demonstrated a high light dark ratio of 565.5 at -2V. It is the result of removing more organic ligands on the surface of PbS QDs and introducing larger amount of I^-. This work demonstrates that the hybrid I^- passivated process has great potential in the performance improvement of PbS QDs based photovoltaic device.

1. Introduction

PbS QDs have attracted increasing interest due to their tunable narrow band-gap (0.4-2.0eV), a large Bohr radius (18nm), a high light absorption coefficient ($\approx 10^6 M^{-1}cm^{-1}$)[1] as well as facile solution-processed. Nowadays, Some of the characteristics have become a new generation of low-cost photoelectric materials[2], such as solar cells[3], photodetectors[4] and light-emitting diodes[5] had been widely studied in the past few years.

The structure of the PbS QDs ($PbS[Pb(OA)_2]$) is composed of a PbS core with a stoichiometric ratio and a Pb shell wrapped around it. The mechanism of passivation is to replace the long chain of organic ligands bound to the Pb shell on the basis of protecting it from oxidation, in order to reduce the interparticle spacing to improve carrier mobility in the PbS QDs film. One method is called atomic ligand passivation[6,7], which the use of halide passivation (such as Br^-, Cl^-, and I^-) to exchange oleic acid (OA) ligands on the surface of PbS QDs, and passivate trap states that lower the minority-carrier lifetime[8], and their thin film carrier mobility reaches $10^{-2}cm^2V^{-1}S^{-1}$[9]. The other method is the organic ligands (such as EDT, EDT, MPA, etc.) replaces the long chain of the PbS QDs film, combines with Pb^{2+}[10-12]. Successfully, Ryan W. Crisp[11]prepared PbS QDs solar cells using metal halide (PbI_2, $PbCl_2$, CdI_2 or $CdCl_2$) salt replaced the oleic acid surface ligand. In recent years report have shown that the surface electronic properties of PbS QDs can be modified by surface ligand chemistry[13]. By replacing the long chains of insulated hydrocarbon ligands to enhance the electronic coupling and carrier transport capacity of the PbS QDs[14].

Due to PbS QDs have strong quantum confinement effects and potential multi-exciton generation effects, which provide the possibility to improve the efficiency of photoelectric devices, and as active layer of PbS QDs materials have a great impact on device performance. Therefore, the paper is devoted to studying the influence of the different passivation methods on the surface composition and optical properties of PbS QDs films, and prepared PbS QDs photovoltaic devices.

2. Experiment

2.1 MAI-treated PbS QDs (PbS_{MAI})
The synthesis of PbS QDs is referred to the previous work from Hines[15,16] with certain modifications. For a MAI treatment process, 1.0 mL toluene, 0.5 mL DMF dispersed 4 mg MAI, the mixture solution was added to 2 mL PbS QDs in a vial tardily (1 drop per 10 s). Gentle shaking was applied to facilitate the formation of homogeneous solution. Note that quickly injection of MAI solution resulted in non-reversible precipitation of QDs. The QDs solution was kept in the glove box for 36 h. Toluene and acetone was employed to precipitate the QDs. Following fully drying, the PbS QDs were dispersed in octane with a concentration of 50 mg/mL.

2.2 Hybrid Passivation (PbS_{MAI}-MAI)
As shown in Fig.1, the oleic acid-capped PbS QDs solution was treated with an MAI (PbS_{MAI}) aqueous solution to prepare a 50 mg/ml solution of a PbS_{MAI} QDs in octane at a low speed of 300 rpm/s 5 s, high speed 3000 rpm/s and spin for 30 s on a clean Si wafer or glass substrate, spin with 10 mg/ml and MAI methanol solution coated on PbS QDs film, and finally washed with acetonitrile.

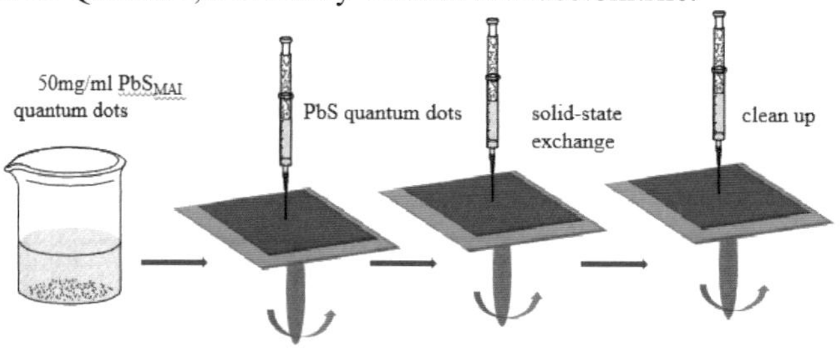

Figure 1. Schematic diagram of hybrid processing of PbS QDs films

2.3 Device Fabrication
The ITO substrates were pre-cleaned in an ultrasonic bath with detergent, pure water, and 2-propanol, acetone, and alcohol for 15 min to remove organic residuals. A thin layer of PbS QDs was formed by spin coating at 300rpm for 5s, 3000rpm for 30s. The PbS QDs film was further annealed at 100 °C for 10min to dry organic residuals and promote crystallization of the PbS QDs film. Finally, samples were transferred to the evaporation chamber for metal contact deposition. 120nm Al was thermally evaporated at a pressure $<5\times10^{-3}$ Pa.

2.4 Experimental Equipment
The X-ray diffraction (XRD-7000) from Shimadzu Corporation was used to test the phase and crystal structure of PbS QDs in the experiment. Ultraviolet and Visible Spectrophotometer (UV-VIS, PE Lambda950) can be used to estimate the size and band gap of PbS QDs. The Fourier transform infrared spectroscopy (FTIR, HIMADZU FTIR-8400S) measurements were carried out. The surface elements and chemical states of PbS QDs film were examined using X-ray photoelectron spectroscopy (XPS, Thermo Fisher K-Alpha) and all binding energies were calibrated by setting the C1s signal to 285 eV. The current

density-voltage characteristics (J-V) curve were measured with the KEITHLEY-2600B equipment manufactured by Tektronix under ambient conditions. The photoelectric devices of PbS QDs were fabricated by thermal evaporating Al electrodes (120 nm, base pressure < 5.0×10^{-3} Pa) onto ITO substrates with spin-coated PbS QDs films. All optical measurements were performed at room temperature.

3. Result and Discussion

Fig.2(a) shows that the XRD spectra of the PbS QDs film passivated with MAI solution-phase and hybrid passivation method. The diffraction peaks observed at 2θ of 25.96°, 29.50°, 43.05°, 50.95°, 62.49°, 70.28°, 78.92°, 84.79° correspond to (111), (200), (220), (311), (400), (420), (422), (511) planes respectively which are good agreement with the JCPDFNo.5-0592. It has been concluded that the deposited PbS QDs flims are polycrystalline in nature with cubic structure. In Fig.2(b) shown that the first exciton absorption peaks of PbS_{MAI} are located at 1033 nm. Compared to the untreated PbS QDs, the absorption spectra are red shifted by 8nm. The red shifts of approximately 9.3 mev and of the PbS_{MAI} films in optical transitions due to electronic coupling reduced interparticle distance, which results in the delocalization of the electrons and holes wavefunctions, combined with the effect of the increased dielectric constant[6]. The red shift of exciton peaks in the PbS QDs film also indirectly indicates the reduction of the distance between QDs particles[17].

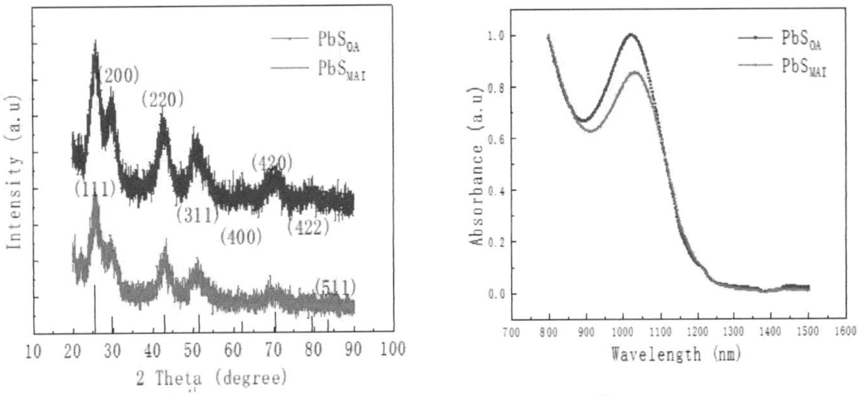

(a) (b)

Figure 2. (a), (b) XRD and UV-VIS spectra of PbS QDs films in PbS_{OA} and PbS_{MAI} Solution processing

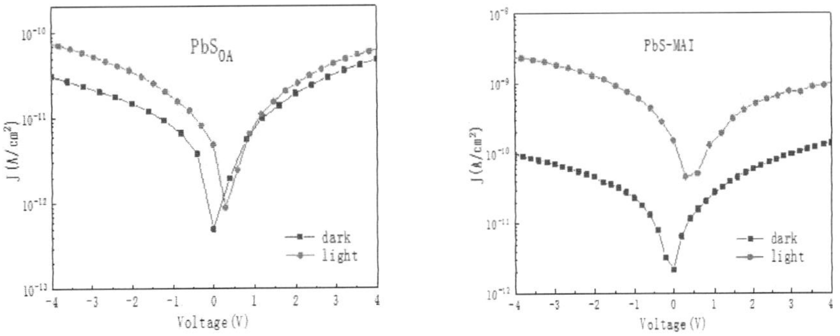

(a) (b)

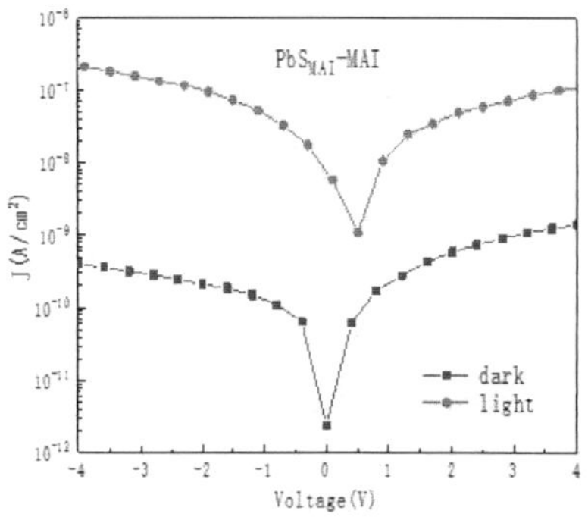

(c)

Figure 3. Current-voltage (J-V) characteristic of devices. (a), (b), (c) PbS_{OA}, PbS_{MAI}, PbS_{MAI}-MAI, respectively

In order to further study the effect of passivation of different materials on PbS QDs photoelectric devices, a device of ITO/PbS/Al structure was prepared. As shown in Fig.3(a), (b), (c), it is the current density-voltage characteristics (J-V) curve of the untreated PbS_{OA}, PbS_{MAI} and PbS_{MAI}-MAI QDs devices, respectively. It can be observed that the untreated PbS_{OA} QDs device has almost no photoelectric effect, and the low of the light and dark current. Under the bias voltage of -4V, the light and dark current density of the PbS_{MAI} QDs device is relatively small and no significant change.The photocurrent has a slight change, which may be due to a large amount of organic solvents remain on the surface of the PbS QDs affects the carrier transport. For hybrid passivation of PbS_{MAI}-MAI QDs devices, the devices have a large light-dark current ratio of 565.5 at -2V, and the photocurrent density has increased by four orders of magnitude.

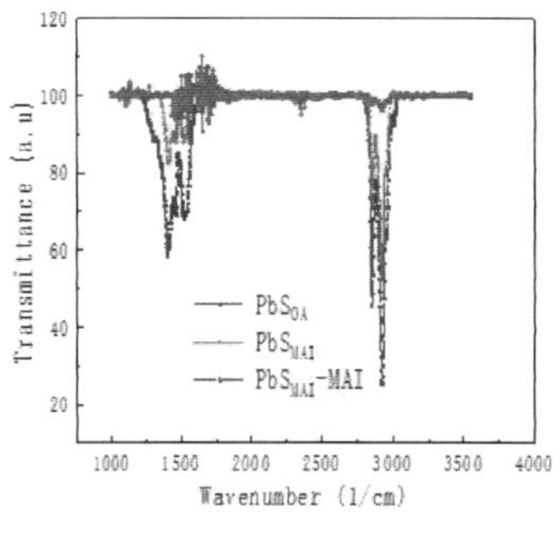

Figure 4. FTIR spectra of PbS_{OA}, PbS_{MAI} and PbS_{MAI}-MAI QDs

In order to investigate the reasons for the improved performance of PbS QDs optoelectronic devices, and to further study the effect of iodine introduction on the optical properties of PbS QDs. The FTIR spectrum in the Fig.4 shown that the vibration intensity of COO-asymmetric and symmetric stretching peaks at 1398cm^{-1} and 1528cm^{-1} decreased, and the vibration peak intensity of CH(-CH_2) at 2852cm^{-1} and 2922cm^{-1} decreased. It can be seen that PbS_{MAI}-MAI QDs can remove of the oleic acid ligands, and it is better at removing oleic acid ligands.

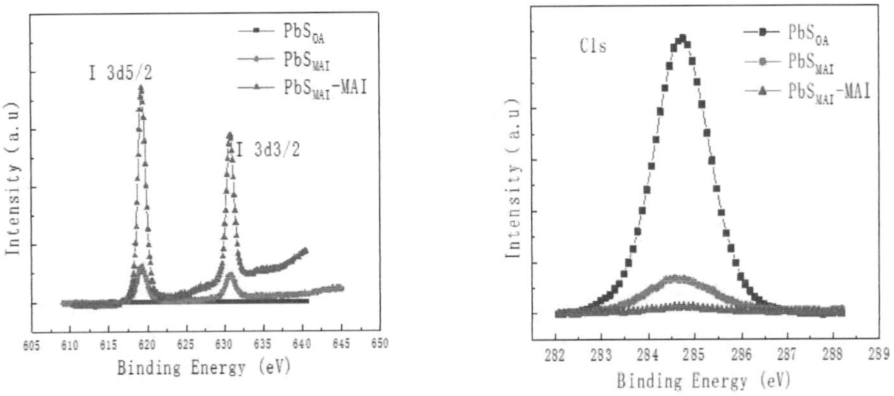

(a) (b)

Figure 5. (a), (b) Comparison of I3d, C1s spectra of PbS quantum dot films passivated by two methods (normalized)

As shown in Fig.5(a), for the PbS_{OA}, PbS_{MAI} and PbS_{MAI}-MAI films the I3d signal with the binding energy at 619.5 eV (I3d5/2) and 632.3 eV (I3d3/2), combined with Pb^{2+} to form PbI_2 and further wrapped on PbS QDs, clearly demonstrate the presence of iodine on the PbS_{MAI}, PbS_{MAI}-MAI films. Similarly, Fig.5(b) it observed that the C1s signal intensity was lower in the original OA ligand. The study found that the PbS QDs by hybrid passivation method is more effective than single solution-phase in removing oleic acid, and capped the surface of PbS QDs by iodine. The strategies has therefore been devoted to minimize the interparticle spacing to promote carrier transport and lower the defect density to reduce recombination loss in the PbS QDs photovoltaic device.

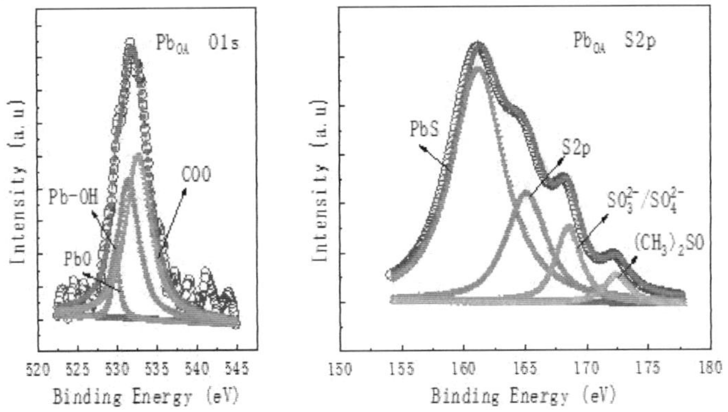

(a) PbS_{OA} O1s (b) PbS_{OA} S2p

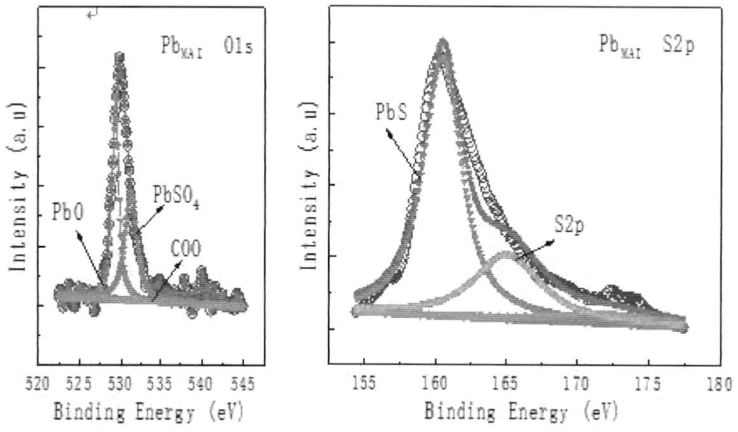

(c) PbS$_{MAI}$ O1s (d) PbS$_{MAI}$ S2p

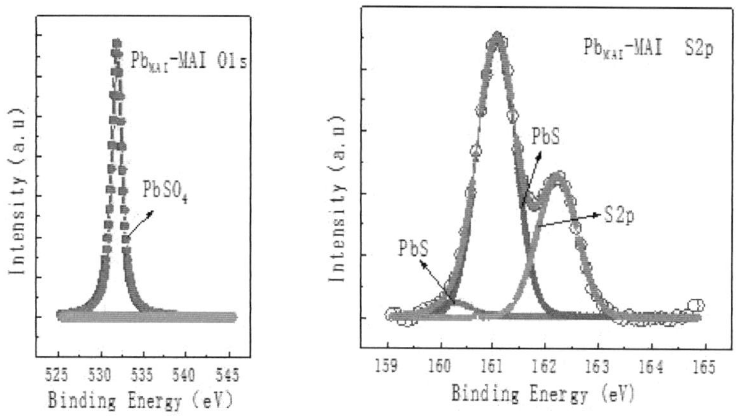

(e) PbS$_{MAI}$-MAI O1s (f) PbS$_{MAI}$-MAI S2p (normalized)

Figure 6. O1s and S2p spectra in XPS of PbS QDs films.

To determine the surface states and chemical compositions of the treatment PbS QDs films, XPS were performed and the results are shown in Fig.6. For the PbS$_{OA}$, PbS$_{MAI}$, and PbS$_{MAI}$-MAI QDs films, the O1s signal with binding energy peaks at 529.3eV (Pb-O), 531.5eV (SO_4^{2-}), and 532.2eV (COO-), respectively. The signal of O element was significantly reduced and increased PbO signal intensity in PbS QDs treated by MAI solution-phase method due to PbS QDs exposure under air environment. For the PbS$_{OA}$, PbS$_{MAI}$, and PbS$_{MAI}$-MAI QDs films the S2p signal (Fig.6(b), (d), (f)) with binding energy peaks at 160.9 eV (Pb-S), 165.1eV (S^{2+}), 168.5eV (SO_3^{2-}(SO_4^{2-})), 172.2eV (($(CH_3)_2SO$)), respectively. It can be seen that the S element signal is slightly reduced, due to the effective reduction of $(CH_3)_2SO$, SO_3^{2-}(SO_4^{2-}) and the partial replacement of the S element by iodine ions, which forms PbI_2. Byproducts such as $PbSO_3$ or $PbSO_4$ are formed in the air environment, which will also reduce the performance of PbS QDs photovoltaic devices[3,4]. For PbS quantum dots with a band gap of about 1.3 eV, the trap state introduced by $PbSO_3$ with an energy level of 0.1 eV is relatively shallow, which mainly plays the role of extending the effective carrier lifetime. And band gap of about 0.8 eV, the trap state introduced by $PbSO_4$ at an energy level of 0.3 eV is very deep, which acts as a recombination center. It can be known from XPS that MAI solution-phase passivation and hybrid I^- passivated of PbS QDs can effectively replace oleic acid ligand and form

$PbS(PbI_2)$ QDs wrapped with iodine ions, which effectively shortens the PbS QDs spacing. Comparing to the O1s and S2p spectrum, it is found that the best way to passivate PbS QDs is hybrid passivation method. Improving the carrier transport capacity will lay a certain foundation for improving the performance of PbS QDs photoelectric devices.

4. Conclusion

The paper focuses on the different passivation methods to treatment PbS QDs, including the solution-phase method and hybrid passivation method. The surface chemical composition, PbS QDs film quality, and optical properties of PbS QDs were studied. The study found that the two processes can remove organic ligands on the surface of PbS QDs, and the most obvious effect is the hybrid passivation method. The ITO/PbS QDs/Al structure with hybrid I^- passivation have demonstrated a high light dark ratio of 565.5 at -2V. The organic ligands on the surface of PbS QDs are almost completely removed, and introduced a large amount of I ions, at the same time the surface forms $PbS(PbI_2)$ QDs, which reduces the dangling bonds or unsaturated bonds on the surface of the PbS QDs. It is further reducing the interparticle distance and increased the coupling energy between the quantum dots, which speeds up hopping and tunneling mechanism. The transport ability has laid a certain foundation for the improvement of the performance of PbS QDs photoelectric devices.

Acknowledgment

This work was supported in part by National Natural Science Foundation of China under Grant 62004161, in part by Natural Science Basic Research Plan in Shaanxi Province of China under Grant 2020JQ-636, in part by Scientific Research Project of Education Department of Shaanxi Province of China under Grant 20JK0796, and in part by Youth Talent Lift Project of Xi'an Science and Technology Association under Grant 095920201318.

5. References

[1] Zhang J, Crisp R W, Gao J, et al. Synthetic conditions for high accuracy size control of PbS Quantum Dots[J]. Journal of Physical Chemistry Letters, 2015, 6(10):1830-1833.

[2] Ren Z, Sun J, Li H, et al. Bilayer PbS quantum dots for high performance photodetectors[J]. Advanced Materials, 2017, 29(33).

[3] Lu K, Wang Y, Liu Z, et al. High-efficiency PbS quantum-dot solar cells with greatly simplified fabrication processing via "solvent-curing"[J]. Advanced Materials, 2018, 30(25):e1707572.

[4] Wei Y, Ren Z, Zhang A, et al. Hybrid organic/PbS quantum dot bilayer photodetector with low dark current and high detectivity[J]. Advanced Functional Materials, 2018, 28(11):1706690.

[5] Wang L, Lv Y, Lin J, et al. High-efficiency inverted quantum dot light-emitting diodes with enhanced hole injection.[J]. Nanoscale, 2017, 9(20).

[6] Tang Jiang, Kemp, Kyle W, Hoogland, Sjoerd, et al. Colloidal-quantum-dot photovoltaics using atomic-ligand passivation[J]. Nature Materials, 2011, 10(10):765-771.

[7] Peng L, Tang J, Zhu M. Recent development in colloidal quantum dots photovoltaics[J]. Frontiers of Optoelectronics, 2012, 5(4):358-370.

[8] Tang J. Materials engineering for stable and efficient PbS colloidal quantum dot photovoltaics[Doctoral dissertation]. Toronto: Unerisity of Toronto, 2011.

[9] Ning Z, Zhitomirsky D, Adinolfi V, et al. Graded doping for enhanced colloidal quantum dot photovoltaics[J]. Advanced Materials, 2013, 25(12):1719-23.

[10] Ip A H, Thon S M, Hoogland S, et al. Hybrid passivated colloidal quantum dot solids[J]. Nature Nanotechnology, 2012, 7(9):577.

Journal of Physics: Conference Series **2065** (2021) 012018 doi:10.1088/1742-6596/2065/1/012018

[11] Crisp R W, Kroupa D M, Marshall A R, et al. Metal halide solid-state surface treatment for high efficiency PbS and PbSe QD solar cells[J]. Scientific Reports, 2015, 5:9945.

[12] Koleilat G I, Levina L, Shukla H, et al. Efficient, stable infrared photovoltaics based on solution-cast colloidal quantum dots.[J]. Acs Nano, 2008, 2(5):833.

[13] Crisp R W, Kroupa D M, Marshall A R, et al. Metal Halide Solid-State Surface Treatment for High Efficiency PbS and PbSe QD Solar Cells[J]. Scientific Reports, 2015, 5:9945.

[14] Brown, Patrick R, Kim, Donghun, Lunt, Richard R, et al. Energy level modification in lead sulfide quantum dot thin films through ligand exchange[J]. Acs Nano, 2014, 8(6):5863-5872.

[15] Hines M A, Scholes G D. M. A, Hines, G. D, et al. Colloidal PbS nanocrystals with size-tunable near-infrared emission: observation of post-synthesis self-narrowing of the particle size distribution[J]. Advanced Materials, 2010, 15(21):1844-1849.

[16] Lee J S, Kovalenko M V, Huang J, et al. Band-like transport, high electron mobility and high photoconductivity in all-inorganic nanocrystal arrays[J]. Nature Nanotechnology, 2011, 6(6):348-352.

[17] Luther J M, Law M, Song Q, et al. Structural, optical, and electrical properties of self-assembled films of PbSe nanocrystals treated with 1,2-ethanedithiol.[J]. Acs Nano, 2008, 2(2):271-280.

Journal of Physics: Conference Series **2065** (2021) 012019 doi:10.1088/1742-6596/2065/1/012019

Influence of Noise on Fault Diagnosis of Transformer based on Vibration Signal

Ming-hao Chen[1], Quan Zhou[1,*] and Yangxi Ou[1]

[1]State Key Laboratory of Power Transmission Equipment & System Security and New Technology Chongqing University, Chongqing, China

zhouquan@cqu.edu.cn

Abstract. Monitoring transformer vibration signals is a universal application method to realize the diagnosis of internal mechanical faults of transformers. However, the actual transformer operating is interfered by the noise of the surrounding electrical equipment, which reduces the accuracy of the vibration signal identification. This paper simulate the typical noise sources in the actual transformer operating environment, including fan noise and surrounding equipment fault noise, and explore the impact of different noise sources on the transformer vibration signal .

1. Introduction

As one of the methods of transformer fault diagnosis, vibration method has become a reliable transformer fault diagnosis technology that has developed rapidly in recent years [1-4]. The vibration signal of the transformer contains fault information of the transformer, but the transformer would also be interfered by the surrounding power equipment and noise signals in actual operation, which reduces the accuracy of the identification of the vibration signal. When the transformer is in operation, the main noise interference comes from the noise interference of its own cooling equipment [5, 6]. Reference [5] compares the frequency spectrum of the vibration signal when the fan is working and the frequency spectrum of the vibration signal when the fan is turned off in the actual transformer operation. The result shows that when the cooling fan is working, the vibration amplitude measured on the surface of the transformer box is There is a significant increase in the low-frequency area within 100z, which causes greater interference to signal processing and analysis. Therefore, it is of engineering significance to explore the interference influence law of multi-source environmental noise on transformer vibration signal and propose a countermeasure. This paper simulates the typical noise sources in the actual transformer operating environment and applies them to the transformer mechanical defect simulation test platform to obtain the vibration characteristics under different noise backgrounds and explore the influence of different noise sources on the vibration signal.

2. Transformer vibration mechanism

Transformer vibration is caused by the vibration of its internal iron core, windings and cooling device. At the same time, external environmental noise will also be transmitted to the surface of the transformer through the air to produce certain vibrations. The vibration and noise generation and propagation process are shown in Figure 1.

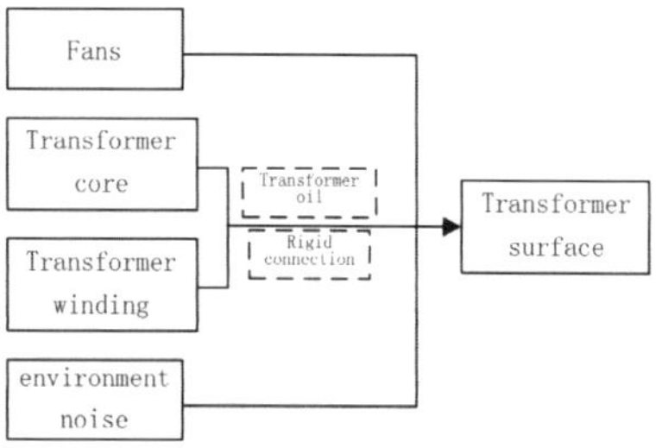

Figure 1. Schematic diagram of transformer vibration and noise generation and propagation

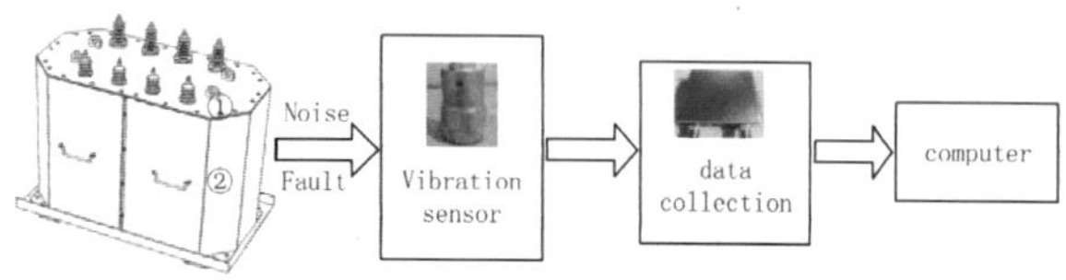

Figure 2. Description of transformer vibration test system under noise

The winding vibration is mainly caused by the dynamic electromagnetic force of the coil passing the alternating current in the leakage magnetic field, and the vibration of the iron core is mainly caused by the magnetostriction phenomenon of the silicon steel sheet and the electromagnetic force caused by the eddy current between the silicon steel sheets. The rotating frequency of the fan has a serious interference to the vibration signal of the transformer, and the working frequency of the fan is affected by the frequency of the power grid and mainly works in the low frequency area . Therefore, the superimposition effect produced causes the change of the overall frequency amplitude.

3. Feature extraction based on vibration signal

This paper first measures the vibration signals of the transformer in various operating states under different noise environments. Obtain the frequency domain diagram of the vibration signal by fast Fourier transform, and analyze the effect of noise on vibration signals.

The transformer vibration test system is shown in Figure 2, including a noise generating device, a piezoelectric acceleration sensor, a signal conditioner, a data acquisition card and an upper computer. The vibration sensor is fixed on the transformer shell through the magnetic base to obtain the vibration signal of the transformer, and it is installed at position ① in Figure 2; the interference noise around the transformer is simulated by the fan and audio equipment. The interference noise around the transformer is simulated by the fan and sound, and the noise source is 50cm away from the vibration sensor, installed at position ② in Figure 2. The signal is transmitted to the upper computer through the conditioner and data acquisition card, and then the state of the transformer is evaluated. The test transformer is a three-phase dual-winding transformer, as shown in Figure 3. In the experiment, 75% of the rated current at a frequency of 50 Hz was used to provide excitation for the three-phase transformer core model to obtain vibration characteristics; the reactor was connected in series to the capacitor ground loop to reduce current distortion. The state of the transformer includes three states: normal operation, loose winding fault and loose core fault. Noise is divided into two types: fan noise and equipment fault noise around the transformer. Among them, the fan noise is generated by the transformer fan, and the fault noise of the equipment around the transformer is generated by the audio equipment through collection.

Figure 3. Test transformer

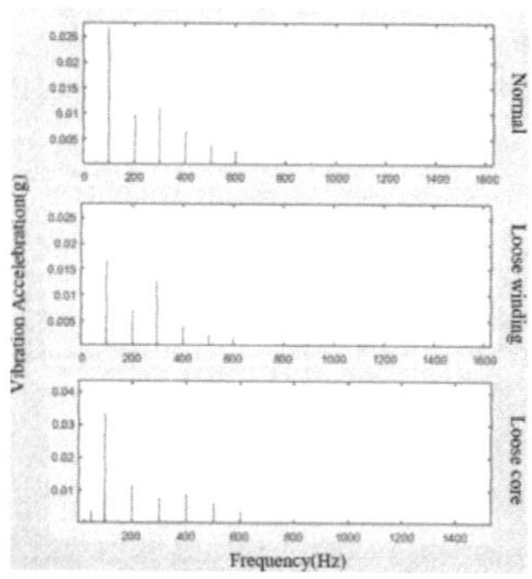

Figure 4. Vibration frequency domain diagram without noise

In a noise-free environment, the vibration signals under the normal state, the loose winding fault state and the loose core fault state were measured respectively, and the frequency domain analysis of the measured signals is shown in Figure 4.

According to the frequency spectrum of the vibration signal calculated by the measurement, the following characteristic quantities can be obtained:

1) Fundamental frequency amplitude:

The fundamental frequency f of the vibration signal is twice the fundamental frequency of the current signal, so the fundamental frequency f of the vibration signal is 100 Hz. After FFT processing, obtain the fundamental frequency amplitude A_{100};

2) Fundamental frequency proportion：

$$P_{100} = A_{100}^2 / \sum_{f=100}^{f_{max}} A_f^2 \quad (1)$$

In formula (1), P_{100} is the fundamental frequency proportion; f is the fundamental frequency of the vibration signal; f_{max} is the maximum value of the fundamental frequency of the vibration signal; A_f is the amplitude corresponding to the frequency f in the vibration spectrum；

3）Vibration entropy：

$$H = \left|\sum_{f=100}^{f_{max}} P_f log_2 P_f\right| \quad (2)$$

P_f is the specific gravity of the vibration signal at frequency f; H is the vibration entropy of the vibration signal at the fundamental frequency f.

4. Analysis of noise interference to signal and fault identification

Under the noise interference of the fan, the vibration signals under the normal state, the loose winding fault state and the loose core fault state were measured. The frequency domain diagram is shown in Figure 5.

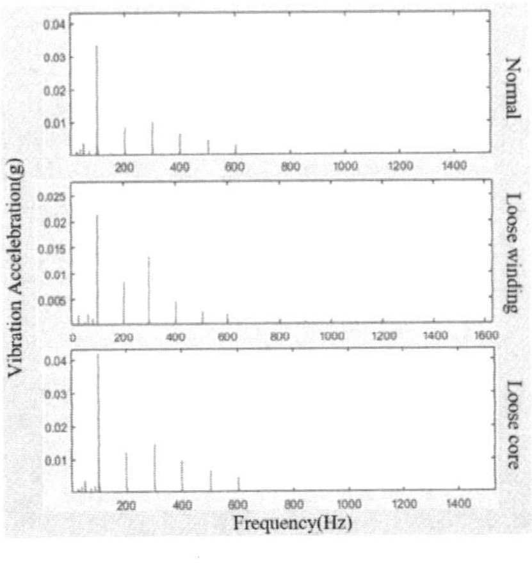

Figure 5. Vibration frequency domain diagram under fan noise

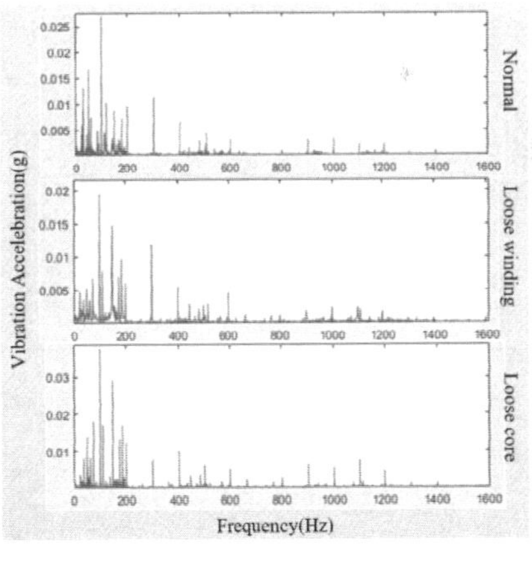

Figure 6. Vibration frequency domain diagram under simulated noise

Comparing Figure 4, it can be found that the fan noise has a greater impact on the low-frequency amplitude of the transformer vibration signal, which is mainly realized as an increase in the fundamental frequency signal. In addition, there is also a small increase in the frequency of integer multiples of 100 Hz. The sound equipment is used to simulate the noise caused by equipment failures around the transformer, and the vibration signals of different states under the interference are measured. The frequency domain diagram is shown in Figure 6. Through the frequency domain diagram, it can be seen that the vibration signal within 1400 Hz will be affected by noise in various frequency bands. By comparison with Figure 4, the frequency band within 200 Hz is greatly affected by noise, and the equipment noise around the transformer is also mainly concentrated in this frequency band. In addition, the fundamental frequency amplitude and the integral multiple of the fundamental frequency amplitude have also increased slightly, but compared to Figure 5, it can be found that the influence of the analog noise on the fundamental frequency amplitude is smaller than that of the fan noise.

According to the frequency domain diagrams of various operating states of the transformer under different noises, the characteristic quantities in each case are calculated, as shown in Table 1. According to the characteristic quantities of the transformer operating in different states under noise, the maximum interference of fan noise to the fundamental frequency amplitude is 38%, while the maximum interference of analog noise to the fundamental frequency amplitude is less than 19%. It shows that the fan has more interference to the fundamental frequency. Fan noise caused a 19% change in the fundamental frequency, while the analog noise had less than 12% of the fundamental frequency, which once again confirmed that the fan has a greater interference to the fundamental frequency. Since the influence of the fan on the frequency is mainly concentrated near 100Hz, and the proportion of the 100Hz frequency is increased, so that the signal frequency domain is more concentrated near 100Hz, so the vibration entropy is reduced. The fault noise of other equipment has an impact on each frequency band of the original signal, resulting in an increase in each frequency band, and a more scattered frequency, which leads to an increase in vibration entropy.

Table 1. Feature in different situations

Noise	State	Fundamental frequency amplitude /g	Fundamental frequency ratio /%	vibration entropy
No noise	Normal	0.0270	72.47	1.35184
	Loose winding fault	0.0160	52.06	1.61092
	Loose core failt	0.0326	76.74	1.25966
Fan	Normal	0.0340	82.24	1.02712
	Loose winding fault	0.0221	62.01	1.51272
	Loose core failt	0.0432	77.34	1.22306
Simulated equipment fault	Normal	0.0271	69.02	1.62164
	Loose winding fault	0.0193	54.44	1.97819
	Loose core failt	0.038	67.54	1.74994

5. Conclusion

This paper analyzes the interference effect of noise on the transformer vibration during the actual operation of the transformer. The interference of fan noise is particularly obvious. Therefore, when measuring the vibration signal of the transformer in practice, the position of the fan should be selected as far away as possible to reduce the influence on the measurement result. At the same time, the environmental noise around the transformer should be monitored, and the operating state of the transformer should be judged in combination with the environmental noise to improve the accuracy of fault diagnosis.

6. References

[1] JI Shengchang , ZHANG Fan , SHI Yuhang , ZHAN Cao , ZHU Yeye. , LU Weifeng, LU Weifeng. Review on vibration-based mechanical condition monitoring in power transformers[J]. High Voltage Engineering, 2020, 46(1): 257-272.

[2] YU Zhangting, LI Dajian, CHEN Liangyuan, et al. Transformer fault diagnosis technique based on voiceprint and vibration[J]. High Voltage Apparatus, 2019, 55(11): 248-254.

[3] WANG S, WANG S, LI H, et al. Mechanical characteristics analysis of defective transformer windings under shortcircuit fault using 3 - D FEM[C]//International Conference on Electrical Machines & Systems(ICEMS). Sydney, NSW, Australia: IEEE, 2017: 1-4.

[4] Su Shiwei.Research on looseness fault of oil-immersed transformer based on vibration analysis method[D]. Huazhong University of Science & Technology,2018.

[5] ZHAO Z Y, TANG C, YAO C G, et al. Improved method to obtain the online impulse frequency response signature of a power transformer by multi scale complex CWT[J]. IEEE Access, 2018, 6: 48934-48945.

[6] WANG Y, PAN J. Comparison of mechanically and electrically excited vibration frequency responses of a small distribution transformer[J]. IEEE Transactions on Power Delivery, 2017, 32(3): 1173-1180.

Journal of Physics: Conference Series **2065** (2021) 012020 doi:10.1088/1742-6596/2065/1/012020

Research on CarMaker and Cloud Computing Platform Co-Simulation Based on APO

Nver Ren, Rong Jiang*, Dongze Zhang

China Automotive Technology and Research Center Co., Ltd. , Tianjin, China

jiangrong@catarc.ac.cn

Abstract. An cloud computing platform based on B/S architecture and docker container technology for autonomous driving simulation has been established in this paper. The map editor module of the cloud platform lets users design 3D scenes for simulating and testing automated driving systems. When the customized roadway scene for simulation created, it would be saved as OpenDrive format both for the server of cloud platform and CarMaker's TestRun which all parameters of the virtual environment (vehicle, road, tires, etc.) are sufficiently defined. Then, based on the application online (APO) communication protocol of CarMaker, the local APO agent service was created. When the 27 parameters of vehicle dynamics received from CarMaker server,they were sent to the cloud platform in real time through UPD protocol. The process of data communication is completed by APO agent. Through the work above, a co-simulation between cloud platform and CarMaker could be successfully established for autonomous driving with seventeen-degree-of-freedom. Through the co-simulation experiment, it is found that the real-time data sampling frequency of the co-simulation is 70Hz, which completes the synchronous simulation of carmaker and cloud platform.

1. Introduction

Carmaker is a vehicle model simulation software based on C / S architecture. At present, there are some co-simulation experiments with CarMaker, such as CarMaker and Simulink were used to realize the modeling and joint simulation for automobile EPS system[1]. A co-simulation between PreScan and CarMaker has been established. The possibility of prescribing the PreScan vehicle motion (wheels excluded) was considered as a verification of the claimed S-function capabiilities and, therefore, of a Prescan/CarMaker co-simulation feasibility[2].It is not difficult to see that most of the co-simulation with CarMakeris based on C/S architecture, which is usually realized by installing multiple clients locally at the same time, and is limited to the version of the software. When we consider using C/S architecture for co-simulation experiments, there are some questions, for example, relatively high cost for development and maintenance,bad distribution,high resource cost when multi software running at the same time,low simulation efficiency and so on. Comparatively speaking, the autonomous driving simulation software based on cloud computing platform can support multi-user parallel simulation[3], besides, it has the following good properties: good distribution and sharing, not restricted by operating system, cross-platform, convenient operation and maintenance and so on. Using the advantages of Docker containers and mirror[4], It proves that the solution proposed in this thesis balance the use of resources of the nodes in cluster and improve the efficiency and stability of the co-simulation[5] experiment for autonomous driving.

With the rapid iteration of autonomous driving technology, the need to test autonomous vehicles has become more urgent. Virtual simulation testing is one of the basic testing technologies which constitute autonomous vehicle testing. In recent years, it has received extensive attention from automotive industry. So, it helps to cover all kinds of test conditions required by combining virtual simulation testing with real-world test[6].Although many companies and scientific research institutions at home and abroad have carried out the research on autonomous driving simulation platform, they are mutually independent and with their own advantages and disadvantages.Traditional simulation software such as CarSim and carmaker are mature in the field of vehicle dynamics simulation. The automatic driving algorithm in the aspect of traffic rules could be evaluated when simulating with Apoll platform. The main advantage of carcraft simulation platform is the verification of automatic driving algorithm. Panosim is mainly used in the development, testing and verification of automatic driving technology. The simulation platform launched by cognata is dedicated to providing highly realistic virtual driving scenarios[7].

In this thesis, a cloud computing platform has been built based on Docker. It is possible to support load scheduling, migration of containers, remote login and continuous delivery. In addition,we can customize roadway scenes by creating region-specific road signs and markings for co-simulation. Finally,based on the APO communication protocol of CarMaker, the local APO agent service was created. When the 27 parameters of vehicle dynamics received from CarMaker server,they were sent to the cloud platform in real time through UPD protocol.

2. Seventeen-degree-of-freedom for Vehicle[8]

To simulate the vehicle motion state under different working conditions, a comprehensive vehicle model must be established. The seventeen-degree-of-freedom vehicle model used in this paper was to consider the steering wheel angle, driving torque and braking torque input. Seventeen degrees of freedom include three axial motion degrees (x, y and z) and three axial rotational degrees of freedom (P, Q and R) of the body mass; The figure 1 shows the seventeen-degree-of-freedom vehicle model diagram, including four vertical motion degrees of mass under the front and rear springs, four spin degrees of freedom of four wheels, steering freedom of two front wheels around the main pin and the freedom of movement of tie rod.

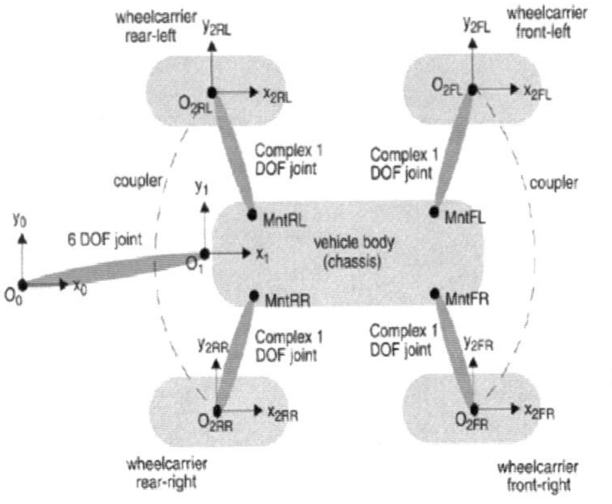

Figure 1. The seventeen-degree-of-freedom vehicle model diagram

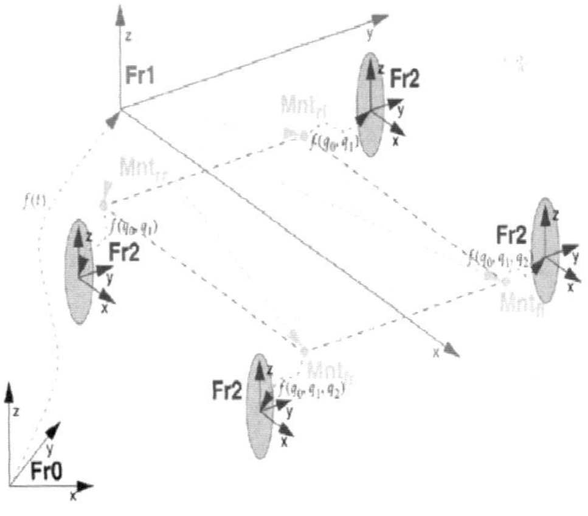

Figure 2. CarMaker coordinate systems

In the virtual world of CarMaker different axis systems for different purposes are used. They are used to simplify calculation and parametrization for CarMaker objects (including signals and variables) and to be able to represent different points of views for CarMaker objects.

In order to describe the motion state of vehicle, body and the force analysis of tire, six reference coordinate systems are established(see Fig.2). The CarMaker inertial axis system is called Fr0 (pronounced: frame zero). This is the earth fixed origin of the 'virtual world'. O is the origin, (O, X, Y) is the horizontal driving plane (road), and Z is directed upwards (mathematically: $X \times Y$).

Journal of Physics: Conference Series **2065** (2021) 012020 doi:10.1088/1742-6596/2065/1/012020

Moving objects in the virtual world are based on their own accompanied axis system which is called Fr1. This axis system is fixed to the moving object. This means that the axis system performs all movements of the attached object like translations and rotations. X-axis positive direction points in forward driving direction, Y-axis positive direction points to the left, and Z is directed upwards (mathematically: $X \times Y$).

For every wheel there is a mountpoint (Mnt) defined within the Fr1 system. This is the center of reference of a Fr2 axis system attached to this mount-point. Mount-points are pure translations (X,Y,Z) from the Fr1 axis system. They are fixed to the Fr1 system.

There are functional dependencies (suspension kinematics and compliance) how Fr2 is orientated relatively to its mount-point. There are three generalized coordinates(q_0, q_1, q_2) for the movement of each Fr2 axis system. Usually q_0 stands for compression and q_2 for steer influence; q_1 stands for compression of the opposite wheel.

3. Map Editor

In order to meet the needs of user-defined driving scene creation, the intelligent driving simulation cloud platform software includes a map editor module. The cloud platform covers the basic elements of the driving scene, such as roads, vehicles, pedestrians, red street lights, roadside buildings and vegetation. As an essential element of driving scene, road has many attributes, such as lane direction, number of lanes, road curvature, intersection, elevation and so on. But in essence, roads can be divided into two types: straight road and curve road[9-10] the specific road modeling process is shown in Figure 3.

Users can click the curve road, spiral road or Island Road button in the function menu, when clicking the left mouse button in the drawing area to generate the road reference line, then the corresponding shape of the road was rendered. After the road was successfully added, the operations of translation, rotation, elevation setting and curvature editing are completed by dragging the control points on road reference line or setting the road property settings window. The map editor also allows users to set the direction, number and width of lanes. After the road related attributes are confirmed, according to the needs of the driving scene, traffic participants such as vehicles, pedestrians, crosswalks, traffic lights, buildings, vegetation and so on are added based on the road reference line to complete the scene generation.

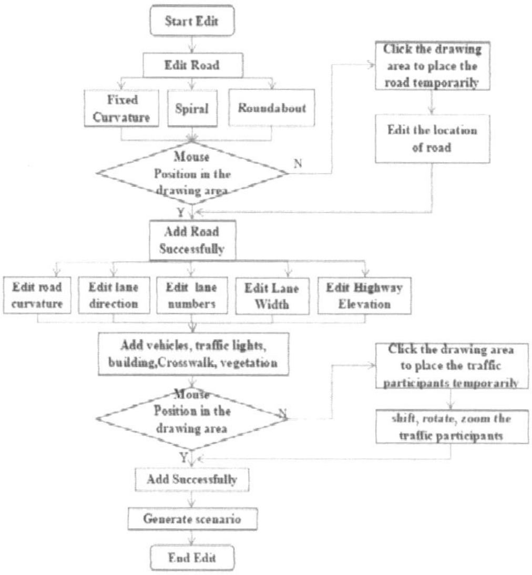

Figure 3. The road modeling process

The map editor module provides basic elements editing, driving scene management and driving scene simulation. The management of driving scene is mainly to manage the roads, vehicles, pedestrians, traffic lights, buildings and vegetation added by users. The simulation of driving scene is able to complete the corresponding simulation experiment of intelligent driving in 3D virtual engine after customizing roadway scenes.

4. APO Agent Service

The APO library is a communications library used to pass messages back and forth between server applications and a client applications. The servers and clients can be running on the same host, on two or more hosts running the same OS and being connected by a network, or on two or more hosts running a different OS and being connected by a network.
The APO library is a higher level library, built over TCP (Stream) and UDP (Datagram) sockets and is intended to provide the user with a simple programming interface, while still maintaining the flexibility required by most time critical applications. APO makes passing messages easy, and eliminates the need to start from scratch each time a server/client communication interface is needed, saving time and countless debugging hours in the process.

The CarMaker application can run on a real time system being connected to one or more clients. The APO library functions are used to pass messages and quantities back and forth between the clients (IPGControl, IPGMovie etc.) residing on the non-realtime host and the CarMaker application running in realtime. This shows a single CarMaker application with one or more clients. It is also possible to have multiple CarMaker applications and multiple clients running simultaneously, since APO is not limited to a single server and multiple clients.

The steps of communication using APO are as follows:

- The first APO function in any client implementation is the call to initialize the APO. Then we should make a subscription and store the quantity pointers and other quantity information that will be used later. Our array has size 27, since we have 27 quantity that represents 17 degrees of freedom of the vehicle.
- Second, query the APO brokers for a list of servers. ApocQueryServers() is first called with a two second timeout. It is important to make sure that the list of servers is used directly after obtaining it, or shortly afterwards. Otherwise the list might be out of date.
- Third, we loop through all the servers in the list. For each server in the list, the Identity field is examined and if the identity is “CoSim”, we have found the correct server. Once the correct server is identified, ApocOpenServer() is called and the server handle(“server id”) is kept in sid. Now that the server handle is known a connection can be made(see Figure 4). ApocConnect() is first called using the server handle returned in the previous loop.
- Finally, 27 vehicle dynamics parameter that were used to describe 17 degrees of freedom of the vehicle were obtained at the frequency of 70Hz per second and saved to the parameter array. Then a UDP connection to cloud platform should be created which would enable us broadcast the received data to the cloud platform. When completing the co simulation experiment, the APO agent service must be closed.

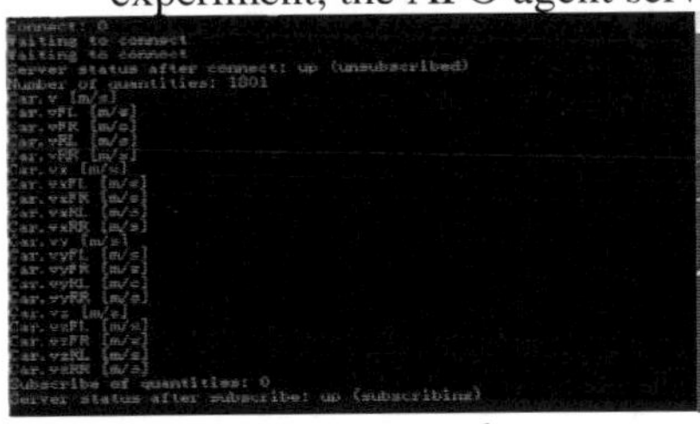

Figure 4. Agent was started successfully to Connect to CarMarker

Figure 5. The synchronization simulation of cloud platform and CarMaker

5. Construction of simulation experiment

The map editor module of the cloud platform lets users design 3D scenes for simulating and testing automated driving systems. When the customized roadway scene for simulation created, it would be saved as OpenDrive format both for the server of cloud platform and CarMaker’s TestRun which all parameters of the virtual environment(vehicle, road, tires, etc.) are sufficiently defined. The synchronization of cloud platform and CarMaker simulation scene is realized, as shown in Figure 5.

The specific steps of co-simulation between CarMaker client and cloud platform are as follows:

- First, log inning to the simulation cloud platform to open the map editing page. After drawing your custom scene, we should save the scene as OpenDrive standard format.
- Second, Start the CarMaker Application and open the scenario editor, the traffic scenario created with the cloud computing platform was directly imported for CarMaker Scenario.
- Second,According to the user's simulation requirements, the initial position and trajectory of the vehicle should be set.
- Third,we should select the experimental vehicle, and edit the driving behavior of the vehicle, including the trigger time of speed, steering, braking and so on.
- Finally, we would start APO agent service, and it helped to send vehicle dynamics parameters to the cloud platform that made the vehicle in Cloud and the car in CarMaker run in synchronization.

6. Conclusion

Virtual simulation testing is one of the basic testing technologies which constitute autonomous vehicle testing. In recent years, it has received extensive attention from automotive industry. Different software has their own advantages, so it is an inevitable trend to realize the comprehensive and efficient experiment of complex driving scenarios by joint simulation. In this paper, the B/S architecture was adopted when developing the simulation cloud computing platform. Furthermore,the advantages of CarMaker application in vehicle dynamics simulation has been made full use of. All the work made it possible to realize the co-simulation of 17 degrees of freedom of vehicles in the cloud platform,which has the advantages of low delay and high reliability.

7. References

[1] Qin Chengxian, Zhang Lei and Yan Ying 2020 CarMaker/Simulink Modeling and Joint Simulation of Automotive ESP System[J]. *Mechanical Science and Technology for Aerospace Engineering*. vol 39, p 1263-1270.

[2] Salvatore Battaglia. Testing the PreScan-CarMaker co-simulation. Tass International Confidential, unpublished.

[3] Su Nianle 2010 Parallelization of Simulation Model Portability Specification on Multi-core Computer[D]. National University of Defense Technology(Changsha, China).

[4] Gao Yuan 2017 Research and Implementation of A Docker-based Private Cloud Host System[D]. University of Electronic Science and Technology of China(Chengdu, China).

[5] Cornetta G, Mateos FJ and Touhafi A et al 2019 Modeling and simulation of a cloud platform for sharing distributed digital fabrication resources[J]. *Computers*. Vol 8, p 47-65.

[6] VISHNUKUMAR H J, BUTTING B and MULLER C et al 2017 Machine learning and deep neural network—Artificial intelligence core for lab and real-world test and validation for ADAS and autonomous vehicle：AI for efficient and quality test and validation[C]. *2017 Intelligent Systems Conference(IntelliSys)*(New York:IEEE), p 714-721.

[7] LIU Jingyu, MA Hui and ZHANG Xuewen et al 2021 Research and development of autonomous driving virtual simulation technology[J]. *CHINA SCIENCEPAPER*. Vol 16, p 571-584.

[8] CarMaker Reference Manual Version 8.0 , unpublished.

[9] Intel open source software technology center 2009 System Virtualization: principle and Implementation Institute of parallel processing, Fudan University *Tsinghua University Press*.

[10] DONALD J. REI FER 2005 Practical Software Reuse. *Mechanical Engineering Press*(China).

Researches on the Technology of Measuring Grain Moisture based on 555 Integrated Circuit

Wan Zhiqiang[1, *], **Zhang Qingqing**[2], **Xu Liang**[3]

[1,2,3]Anhui Technical College Of Mechanical and Electrical Engineering, Wuhu, Anhui, China
[1,2,3]16 Wenjin West Road, higher education park, Yijiang District, Wuhu City, Anhui Province

0120160005@ahcme.edu.cn

Abstract. Grain is the material basis of human existence and development. Food quality has an important effect on people's physical health for our country. Moisture content of grain has been an important factor in affecting food quaility. Every year, a substantial part of the food production is lost because of too much water contained in the food without drying. According to the annual statistics of the country's harvest, loss of food due to drying time caused by water take up 500 - 10000000 tons, according to approximate 1.5% - 3% of the total grain output.With the development of science and technology, grain moisture detection, especially the real-time and accurate detection, has attracted more and more purchase, production, processing, storage, transportation, and attention from various aspects. Under this problem context, hardware-circuit-control system of grain moisture detection based on ATmega128 has been investigated and designed.

1. Introduction

China is the most populous country in the world and is also the most populous country of food consumption. Food and agriculture development has been a national problem in ou country, because of our country's population and the relative scarcity of resources. It is very necessary and important to solve the grain problem with the development of social economy and the growth of population.

China has been the world's largest country of food production and consumption. Each year, there has been about 500 million tons of grain production. The production of rice was about 160 million tons in 2003 among the total output of grain [1][2]. In some areas in our country, especially in the southern region, it often meets a rainy weather in the rough rice-harvesting season because of the temperature, climate, and other reasons. This request will receive from the paddy grain timely drying, to meet the requirements of humidity, in order to better stored grain. It thus is very necessary to design the grain-moisture-measurement system.

2. Introduction of the grain moisture detection

2.1 Capacitance method

Capacitive grain moisture detection method is based on the water content in different grain, grain dielectric constant is also different grain drying and relative dielectric constant is much smaller than that of water, and the water content of grain relative dielectric constant of drying grain and water between the principle to measure the moisture content in grain. Capacitance method has the advantages of simple detection of economy, high reliability, and convenient operation, where the defects are that it is affected by the environment and temperature, and the follow-up data-processing is complex.

2.2 Resistance method

The principle of resistance method is the use of grain moisture content is different, its conductive ability is also the conductivity of different principle to measure grain moisture content method.Resistance and moisture content relation can be expressed as:

$$M = k_1 + k_2 \ln R$$

k_1 and k_2 are constant. Resistance method has the advantages of simple structure and low cost. The deficiency is that measuring result by commissariat breed the impact is bigger, and prior to grain grinding, made according to required size and shape of the sample for the detection.

2.3 Microwave method

Microwave method of grain moisture detection technology, is the use of water on microwave absorption of energy and microwave cavity resonator frequency, phase and other parameters with the moisture variation principle for moisture measurement method [3]. Microwave method has the advantages of high sensitivity, fast speed, and non-destructive measuring sample,where the disadvantages are complicated device and high manufacturing costs.

2.4 Magnetic method

Magnetic method is the high technology content of the method for measuring grain moisture content, the principle is that under certain conditions, whose nuclear spin reorientation, thereby allowing the food at a frequency determined by the absorption of electromagnetic energy, which energy is absorbed and the amount of samples contained in the nucleon number proportional relationship [4][5]. Magnetic method has the advantages of high accuracy, rapid measurement, wide measuring range, and being able to distinguish the bound water and free water. The disadvantage is it leads to high cost and large cost of equipment maitenance.

Grain moisture detection method has many other methods such as infrared method, neutron method, sound method as so on. It has enough space to introduce all methods due to the length of relationship.

3. The 555 integrated circuit in grain moisture detection application

3.1 Detection principle

The grain moisture measurement system has been investigated in this paper. The first is the need to select the core control chip. Then, Atmega128 single chip is used according to the task requirements.

ATmega128 MCU is the AVR series of high-end products made by ATMEL company, which has a low power and high speed, super function, and reduced instruction characteristics. ATmega128 has 133 instructions, of which most instructions can be completed within one clock cycle, and has as many as 32general registers and peripheral control register. This can eliminate common SCM Single cumulative data processing that brings bottleneck phenomenon. Thus, the instruction code is more flexible, more easy coding [6][7].

Capacitance method is used for measuring grain moisture in this paper. The method can save the cost for enterprises, and the hardware and software are relatively easy to achieve. The principle of measurement is that the food to be measured has contained certain moisture, equivalent to the capacitance relative dielectric constant. It can be easy to get the frequency needed with the help of 555 integrated circuit. The frequency can be converted into a corresponding humidity value through the corresponding formula.The result can be displayed by LCD finally.

3.2 Testing circuit diagram

ATmega128 MCU is used as the core of grain moisture measurement control system.Oscillation circuit produced by 555 chip is used for frequency output signal.It can make the system in grain moisture measurement more faster and more efficient executed by using RISC structure of the Atmega 128 chip [8].

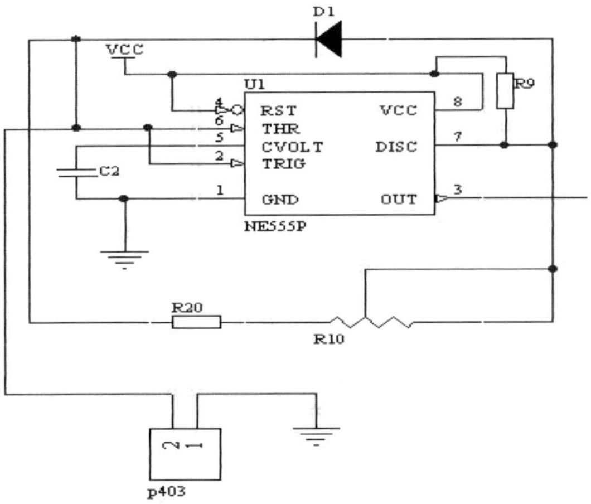

Figure 1 The circuit chart of humidity signal acquisitition

AS is shown in figure 1, where 555 chip output port is connected with the corresponding port3 of Atmega128 MCU. Detection of food frequency signal will be sent to single-chip analysis, where processing. Humidity measurement result value will be displayed on the liquid crystal display finally.

3.3 Program design

The whole process of grain moisture measuring system is that the control system has been connected with the power supply, where the control system will start self-checking and parameter initialization, assigne a corresponding port address, so that the temperature and humidity signal will be acquired respectively.The measured results will be displayed in the LCD. The whole control system of grain moisture measuring instrument will begin to enter the normal working state.The flow chart of system program is shown in figure 2.

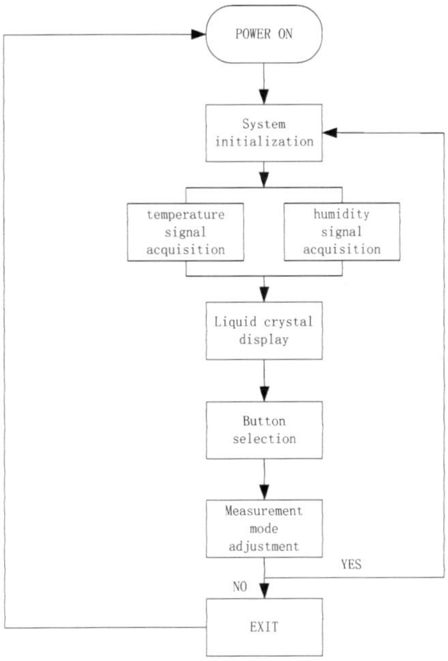

Figure 2 The flow chart of system program

3.4 Insufficient

It will have the advantages of simple structure, low cost being easy to realize the continuous rapid measurement with the application of the principle of grain moisture detection system. The deficiency of this method is that the measurement of the accuracy will not be high enough, and the output frequency will fluctuate with the ambient temperature changing greatly.

3.5 Need for improved technology prospect

Some improving measures will be included in the signal anti-interference circuit in order to improve measured results. The influence of signal input source will be improved in addition to the grain moisture content moisture signal input end.

This can be done by adding a low-pass filter circuit. The low-pass or band-pass filter circuit can be increased in order to get the frequency of the signal accordance with the grain moisture detection of the actual needs. So the measured results can get high precision value.

4. Conclusion

Grain moisture measuring technology is a field of agricultural production that has a very important significance. Some developed countries abroad, such as the former Soviet Union, Japan and other countries began the research earlier. With the grain moisture measuring technology has made great progress, the market has also introduced a lot of good products.In our country,the research in this area is relatively late, Some schools such as Wuhan Polytechnic University, Shenyang University of Technology and Heilongjiang Bayi Agricultural University have done a lot of research on measuring the moisture content of grain. Some domestic scientific research institutes such as Wuhan Kate electronics have done a lot of research in this area at an earlier time, and introduced to the market a lot of convenience and practicality of grain moisture measurement products.

The grain moisture measurement control system investigated in this paper has been in the full reference of the existing research results, design, development of a novel, energy-saving and environmental protection type grain moisture measurement products. ATmega128 single chip microcomputer is the core of hardware of grain moisture measuring instrument ,and on this basis the software has been designed and developped.

Modular structure design ideas have been used in this system. Computer serial communication technology has been introduced in order to achieve automation and intelligent operation.

The PROTEL DXP 2004software produced by Altium company has been used to design the system hardware circuit, The software part has adoptted CodeVisionAVR integrated development environment, In the course of system development proteus software has been used for circuit simulation.

5. References

[1] Wang PengJun. The humidity control system based on configuration software. Northeastern University. 2005

[2] Liu ShuiChang. China's rice production and demand [J]. study on the food problem 2005, (3):16-18

[3] Zhang YongLin. Zhang LianQuan. Liu WenSheng. Comments on on-line real-time measurement of grain moisture content.Journal of WuHan polytechnic University 2003

[4] Hu JunGang, Modern NMR technique application in Food Science [J].Food research and development, 2000, 21(1)：11-15

[5] Jin YongBao, Ai YanBao. Nuclear magnetic resonance technology and its application[J]. Physics and engineering, 2002, 12(1)：47-48

[6] Liu LanXiang, Zhang QiuSheng. ATmega 128 microcomputer application and development examples. China Machine Press. 2006.4

[7] Ma Chao. AVR MCU embedded system principle and application practice. Beihang University Press, 2007.

[8] Shen JianLiang, Zhao WenHong, Jia YuKun, Wang HuaDong. The introduction and improvement of ATmega128single chip microcomputer [M]. Bei Jing: Beihang University press, 2009

Topology Optimization and Design for Additive Manufactured Supporting Structure of Vehicle Rudder

Jingyi Chen[1,*], Kehao Xin[1], Xiaolong Gu[1] and Hongjuan Ji[1]

[1]Beijing Institute of Space Long March Vehicle, China Aerospace Science and Technology Co., Beijing,100076, China

*E-mail: chenjy@14.calt.casc

Abstract. The object investigated in this work is a design method for the supporting structure of vehicle rudders, which are the crucial components in flight control system. As a practical application of topology optimization for additive manufacture, the design process was presented in this article. To obtain the optimal material distribution, topology optimization based on SIMP method was applied, with compliance minimization as the optimization objective. A numerical optimization model was built and calculated in Inspire, and the result in an equivalent load scheme of 30 percent point load and 70 percent uniform pressure was adopted to be the initial geometry for the subsequent design. Concentrating on realizing the optimized geometry, this design took a series of technical constrains for additive manufacture and mass requirements into consideration. Notably, the FEM simulation for the supporting structure is necessary, which consist of stress analysis and mode analysis. And the technical feasibility was verified in its practical manufacturing process. Finally, by the experiment about dynamic characteristic, it is verified that the lightweight rudder designed in this work can satisfy all the requirements for flight control.

1. Introduction

As a key component in flight control, traditional rudders of the vehicle commonly consist of the metal frame skins, and insulation composite layers or coating for thermal protection, as fig.1. Generally, to reinforce the integral stiffness, radial and circular ribs, centered on the rudder shaft axis, are used in the design of frames. Considering the mechanical properties in high temperature, the frame is a precision casting of titanium alloy (Ti-alloy). With Ti-alloy skins welded on it, the main supporting structure is similar to the sandwich panel. The manufacture involves a series of complicated processes, such as Investment Casting and Vacuum Brazing, that is a crucial factor to influence the cost of rudders.

With the development of additive manufacture, mechanical properties of Ti-alloy products have been close to Ti-alloy forgings and the precision have been raised above the level of Investment Casting. SLM (Selective Laser Melting), a branch of additive manufacture, may provide a new approach to manufacture the rudder supporting structure, due to its advantage of accommodating highly complex geometries and inner void structures to mold the rudder frame and skins integrally, instead of welding. In particular, SLM provides an opportunity to make components close to their theoretically optimal geometry. In order to generate a robust support-free structure there are some technical constrains which must be satisfied, such as the minimum feature size, the bridging distance, the manufacturable inclination angle [1], the traditional shape of rudder supporting structure has not been completely appropriate for additive manufacture anymore. Consequently, this work focus on a design method for the supporting structure of vehicle rudder with combining topology optimization and practical experience. The proposed method is required to be appropriate for the metal additive manufacture. Besides, due to the design of rudders must meet the strength and stiffness tolerance requirement in flight, the finite-element method is applied to analyze the structure stress, modes and nature frequency that will be verified by the subsequent test.

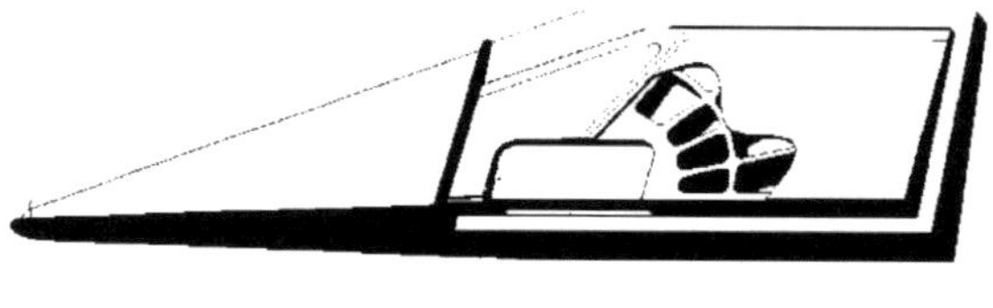

Figure 1. Structure and compositions of traditional vehicle rudder.

2. Topology optimizing model

In this work, SIMP is employed for the rudder supporting structure optimization in order to obtain an adequate strength, high stiffness and light weight geography.

Rossow and Taylor demonstrated the finite element representation of potential energy for the first time, to seek the optimal distribution of material for sheets [2]. Based on the homogenization theory, the continuum structure is discretized into finite units, and the variable (ρ_e)about density is introduced for each unit.

1 or 0 denote the density of the unit is one or void. Ω^{mat} denotes the one material domain, Ω denotes the reference design domain.

$$\rho_e = \begin{cases} 1 & e \in \Omega^{mat} \\ 0 & e \in \Omega \backslash \Omega^{mat} \end{cases} \quad (1)$$

In order to avoid 'grey' material units (intermediate densities), Bendsoe and Zhou [3,4,5], proposed the SIMP (Solid isotropic material with penalization) method. The interpolation scheme and the relation between the density variable (ρ_e) and the young's modulus (E^0) for a unit is represented as follows

$$E = \rho_e^n E^0, n > 1. \quad (2)$$

Where n is defined as the penalty parameter. With intermediate densities (grey) being penalized, the number of 'grey' is reduced effectively, that helps obtain a true "0-1"design. With the finite element method, the minimum compliance problem for elastic structure under loads and the fixed volume conditions is as follows:

$$\begin{aligned} &\text{find } \rho_e, e = 1, \ldots, N \\ &\min c = \mathbf{U}^{\mathrm{T}}\mathbf{K}\mathbf{U} \\ &\text{s.t. } \mathbf{K}(\rho_e)\mathbf{U} = \mathbf{P} \\ &\sum_{e=1}^{N} \rho_e v_e - \overline{V} \leq 0 \\ &0 < \rho_{min} \leq \rho_e \leq 1,\ e = 1, \ldots, N \end{aligned} \quad (3)$$

Here, N denotes the total amount of units, K denotes the stiffness matrix, U and P denote the compliance and load vector, v_e denotes the volume of the eth cell, and $\overline{V}$ denotes the total amount of material volume.

In this work, the numerical model of topology optimization for supporting structure of vehicle rudder is built in Altair Inspire, which is a FEM analysis and optimization platform with rapid simulation capabilities. The initial CAD model consist of external composite layers and the internal metal supporting structure, and material properties are shown in Table 1.

To simplify the optimization model of rudder, the composite layer, which does not belong to the design target, is simplified as an isotropic material in the preprocessing. According to practical experience in rudders design, the joint region between the rudder and shaft, the zone in a distance of

6mm around the rudder edges, and skins are the nonoptimized domain, defined as the external boundaries, other internal areas are the optimized domain, defined as the design space of this model.

Table 1. Material properties of the model.

	Yong's modulus (E)	Poisson radio (μ)	Density (ρ)	Yield strength ($\sigma_{0.2}$)
Titanium alloy (TC4) of SLM	109 GPa	0.34	4.5×10^3kg/m^3	830 MPa
Composite layers	5 GPa	0.07	1.7×10^3 kg/m^3	100 MPa

In this model, thickness constrains are defined to control the wall thickness. The minimum thickness is 3mm, and the maximum thickness is 5mm. In general, thickness constrains also affect the element size, as the minimum thickness is 3 times the average element size.

A Support in all directions is applied on the joint region, to simulate the constrain of shaft. For rudders, normal pressure with variable distribution is the essential load case. Thereby, in this work, the pressure load in the typical flight condition (Table 2) of the vehicle is applied on the rudder surface. However, due to it is not available to import the aerodynamic loads field in Inspire, point loads and uniform pressures of zones have been applied on the surface as an equivalent of the aerodynamic loads in this optimization model. The surface of the rudder is divided into eight zones, and the details about equivalent loads in different zones are shown in Table 3.

Table 2. The typical load of rudder in flight.

Mach Number Ma (/)	Dynamic pressure Q(Pa)	Normal force F(N)	Torque T(Nm)	Bending MM(Nm)
4.19261	730884.50994	-1012.73884	-19.21413	45.78004

Table 3. The equivalent loads and pressure centers of zones.

Section No.	1	2	3	4	5	6	7	8
range of the section(mm)	25.0	70.0	130.0	180.0	236.0	280.0	340.0	410.0
Pressure Center(X)	-10.3	-45.5	-113.2	-139.4	-200.7	-157.4	-171.8	-179.6
Pressure Center(Y)	32.1	51.7	105.0	158.3	211.2	258.2	311.2	374.0
Total normal force of the section	11.5	14.0	23.8	35.0	45.2	55.3	53.1	53.4

3. The optimistic result

Fig. 2 shows the optimal geometric result with equivalent point loads applied on the pressure centers of zones. Fig. 3 shows the optimal geometric result with equivalent uniform pressures applied on zones. Combining the point loads and pressures in proportion, the result with synthesized loads is showed in Fig. 4. In various equivalent schemes, stiffness maximization is still selected to be the optimization objective, while the mass target of 30 percent is specified for the first two cases, and 20 percent for the last.

According to the comparison of these optimal results, the optimal material distributions in various equivalent schemes are extremely close, while some differences is noticeable in detail. Compared to the pure pressure case, the optimization result in the point load case is more sensitive to the location of pressure centers. Furthermore, with the stiffness of the composite layer increasing, the tendency will be more apparent. Therefore, by comparing other results in various proportions of the point load to uniform pressure, the result in 30 percent of point load and 70 percent of uniform pressure is selected to be the basic topological geometry for further design.

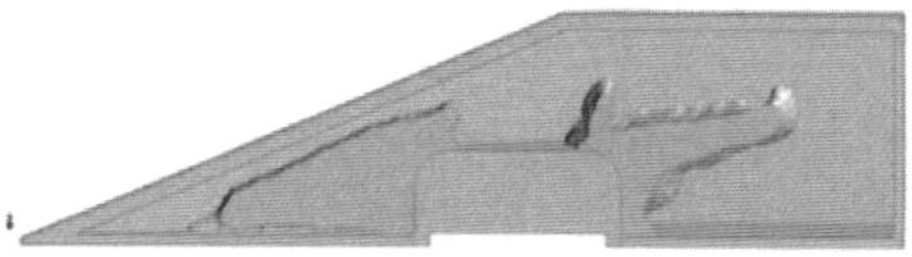

Figure 2. The optimal geometric result with equivalent point loads, 30% mass target.

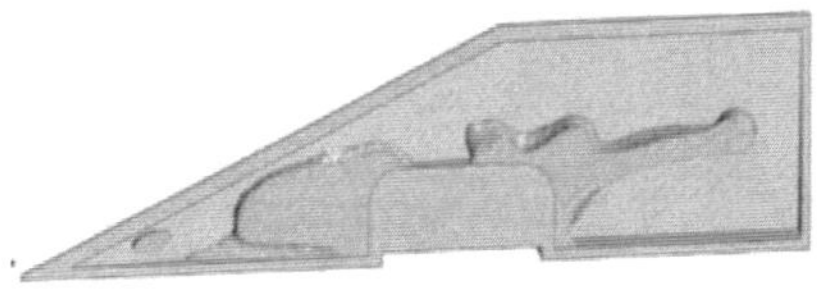

Figure 3. The optimal geometric result with equivalent uniform pressures,30% mass target.

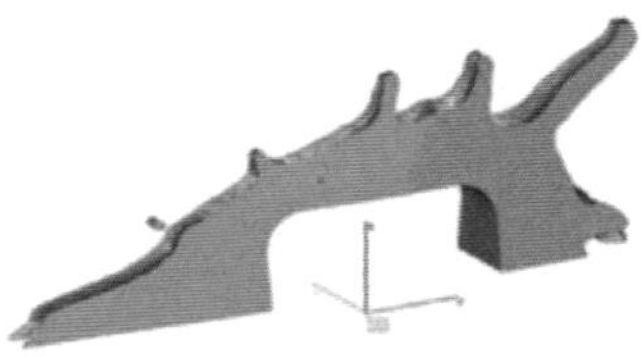

Figure 4. The optimal geometric result with synthesized loads, 20% mass target.

4. Reconstruction and product design

With the benefit of the development of additive manufacture, it is feasible to reconstruct the geometric result of topology optimization with minimum information loss, that due to its less limit in geometry than traditional subtractive manufacture. According to the optimal topology obtained in section 3, the rudder supporting structure is rebuild by the fitting method of NURBS curves and designed in detail. Even if the additive manufacture is labeled as a less limit and dimension free technology, there are a series of constrains must be satisfied in order to improve the feasibility in technology, especially for the closed shape which lead to a support-free design being employed, such as the rudder. For SLM, the factor sought be considered in design are mainly the direction of 3D-print, bridging distance, minimum feature size, and inclination angle. In order to reduce the cost and time in manufacture, the supporting structure was finally built in span direction. The CAD model of the ultimate design is shown in fig. 5, and A Ti-alloy product of the supporting structure of rudder is as fig. 6.

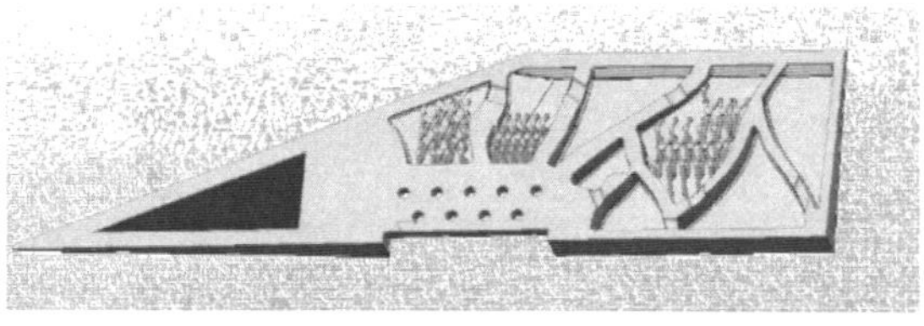

Figure 5. The ultimate CAD model of the rudder supporting structure.

In the design of aerospace vehicle rudder, there is a particular requirement about mass properties of the whole construct: the mass center need to be located at the axis of rudder shaft. Therefor a high-density (Wu-alloy) component was inserted into the tip of the rudder to balance the weight by laser cladding technology. From practical experiences, the inclination angles of ribs are mostly limited in 60 deg. Nevertheless, there are some parts having difficulty to attain it, where three-dimensional lattice structures are used. And as a kind of less density structure, it is not only an approach to remedy the inclination angle problem, but also provides an effective path to improve the local stiffness with less weight. Fig. 7 show the inner 3D lattice strictures of the product by X-rays inspection.

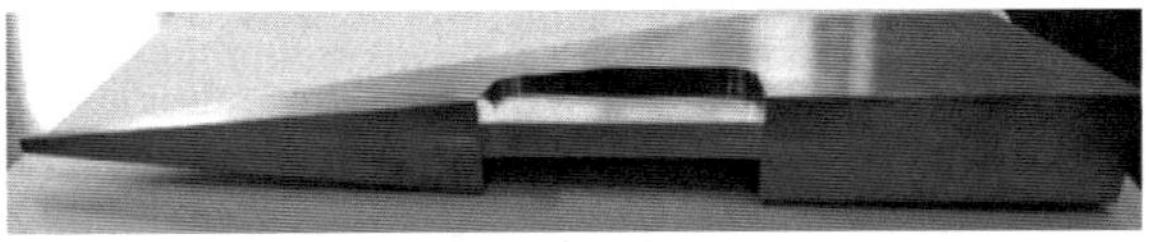

Figure 6. A Ti-alloy product of the supporting structure of rudder.

Figure 7. Inner 3D lattice strictures of the product by X-rays inspection.

5. Stress and mode analysis

In order to verify the strength of the supporting structure of rudder designed in section 4, FE stress analysis by Abaqus is conducted for the entire rudder, with the composite layers included. In the analysis, to avoid the repeat of load cases, maximum normal loads in the extreme flight is applied on the rudder surface in the form of the original pressure field, instead of the equivalent loads. The iso-contours of von Mises stress are plotted in fig.8. With the safety factor of 1.5, the zone of maximum stress is located nearby the connecting section, where the stress is equal to 322MPa. According to the assessment for the strength of whole structure designed above, the remaining strength factor is approximately 2.6, which means that the design in section 4 is still relatively conservative, and there is certain room for further optimization.

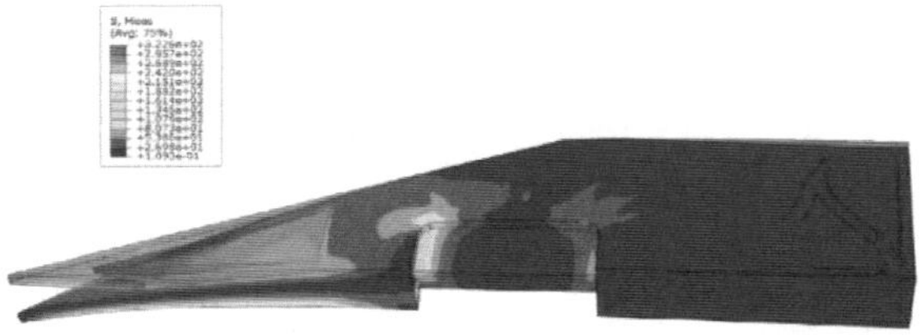

Figure 8. Von Mises stress of the rudder supporting structure.

In the mode analysis, a full-scale assembled model of rudder described above is created with a shaft model merged in, which is used to apply the elastic restraint to simulate the realistic constraints of driving motor and bearings. Two SPRING1 elements with displacement degree of freedom associated are used to simulate the bearings, and a torsion SPRING1 element (associated with rotational degrees of freedom) is used to simulate the driving motor. The result of the mode analysis is listed in Table 4. It is apparent that in comparison with traditional rudders there is a certain improvement in the bending mode frequency of the optimized rudder, which is associated with the stiffness of the supporting structure.

Table 4. The result of mode analysis.

Mode NO.	Mode Frequency (Hz)		Mode shape
	Optimized rudder	Traditional rudder	
1	170.94	173.54	Torsion
2	290.87	266.05	Bending

6. Experimental test

In order to verify the optimized rudder designed above can meet the requirement of control system, a dynamic characteristic experimental test has been finished. With a real vehicle cabin and servo motor as support boundaries and controller, the resonant frequency and transfer function of the flight control system were revised through the test. To obtain low order modes of the rudder for further flutter analysis, 11 displacement sensors were attached on rudder surface, and a vibration exciter was used to output the vibration force. Furthermore, for reducing the influence of gap nonlinearity, a normal load of 30kg was applied on the tip of rudder. The mode parameters obtained in this test have been list in Table 5, and the mode shapes are presented in fig.9. Compared with the result of the mode analysis above, it shows a consistency in the shape of torsion mode. However, the torsion mode frequency in the test is relatively lower than the value in calculation, due to the difference of brace stiffness between actual boundaries and Spring elements in the simulation, which could be reduced through revising the spring stiffness associated or defining Dashpot elements. For the bending mode, the shape and frequency are both approximate to the results of the FEM simulation in section V, which depends on the individual structural stiffness of the rudder more.

Table 5. Mode parameters in the test.

Mode NO.	Mode Frequency (Hz)	Damping ratio (%)	Mode mess (kg)	Load (kg)
1	138.3	7.68	2.75	15kg
2	313.5	5.30	2.00	

a) Torsion mode b) Bending mode

Figure 9. Mode shapes obtained in the test.

7. Conclusions

The application of topology optimization method in the design of vehicle rudder supporting structure has been presented in this article. With SIMP method, a high stiffness material distribution has been achieved. Based on the optimal geometry, a manufacturable CAD model has been rebuilt, and accomplished efficiently by additive manufacture, with a series of associated requirements to be satisfied, such as mass center location and geometric tolerances. This design is verified in the FEM analysis and dynamic characters experiment. The method could also be transferred to other rudders and wings. The conclusions are as follows:

a) In topology optimization with FEM software, it is turn out to be an effective approach to optimize with a typical load case and verify with extreme load case, considering the dynamic characters.
b) Due to optimal results are sensible to loads, original pressure distribution is still the best choice for topology optimizations of rudder supporting structures, even if the load approximation was employed in this work. And the influence of load schemes and thickness constrains on optimization results should be paid attention to.
c) With SLM technology, the use of 3D lattice structure could significantly improve the local stiffness with minimal increase in weight, especially for the large-sized inner void region.

Furthermore, a secondary topology optimization for this optimized design model may lead to a more balanced result, if required.

Acknowledgment

The author would like to thank the Xian Bright Laser Technologies Company for its assistance in this study.

References

[1] Optimal topology for additive manufacture: a method for enabling additive manufacture of support-free optimal structures[J]. Materials and Design, 63(2014), 678-690
[2] Rossow M P, Taylor J E. A finite element method for the optimal design of variable thickness sheets [J]. AIAA, 1973, 11, 1556-1569

[3] Bendsoe M P, Sigmund O. Material interpolation schemes in topology optimization[J]. Archive of Applied Mechanics, 1999, 69(9), 635-654

[4] Rozvany G I N, Zhou M, Birker T. Generalized shape optimization without homogenization[J]. Structural and Multidisciplinary Optimization, 1992, 4(3), 250-252

[5] Rietz A. Sufficiency of a finite exponent in SIMP (power law) method[J]. Structural and Multidisciplinary Optimization, 2001, 21(2), 159-163.Electronic Publication: Digital Object Identifiers (DOIs)

Analysis of the Response of a Static Reactive Power Compensator to Instability and Failure in Electrical grids

Víctor Manuel Maridueña[1*], Edwin Arnaldo Castro[1], and Nelson Layedra[1]

[1] Universidad Politécnica Salesiana /Electrical Department, Guayaquil, Ecuador

Email: vmariduena@est.ups.edu.ec

Abstract. The reliability of electrical power systems has led to the implementation of new equipment with reliable technology to solve transient failures, in recent decades flexible AC transmission systems (FACTS) have been implemented in power grids, resulting in high levels of stability and control. One of the elements used is static VAR compensators (SVC), however there is very little information about the dynamic response of the device to network instability and electrical failures, for which Simulink analyses the response of the SVC. The device consists of a 47.1 MVar reactive power compensator and a 97.6 MVar inductive reactivator compensator, implemented in a three-phase 500 kV system. The results indicate the effectiveness of response against network instability while maintaining the stable voltage of the network, but against electrical failures the type and time of failure must be considered. In the case of phase-phase faults, the response of the SVC is limited with drops of 0.52 pu.

1. Introduction

Your goal is to simulate the usual appearance of papers in the. We are requesting that you follow these guidelines as closely as possible. The development of populations is directly linked to the reliable supply of electricity, in recent years power electrical systems have had an accelerated change in the rate and growth of demand due to the change in consumer habits, therefore several inconveniences in the operation and quality of service are generated due to the approach to stable operating limits, so it is necessary to incorporate new technologies to maintain the quality of supply.

New technologies seek to improve the transmission and flexibility of electrical power systems. Flexible Transmission Systems in Alternating Current (FACTS) are devices with the ability to control various network parameters seeking to meet two objectives, the first is to increase the transfer capacity of the power system and maintain the operating conditions of the power system by flowing the network paths. The operation of FACTS improves, controls and makes the operation of PMI [1]–[3] improved, controlled and more flexible.

One of the widely used devices in the FACTS family is static reagent compensator (SVC), its main function is to provide control of the magnitude of the bus voltage, regulates the bus voltage to compensate when you have reactive power changes, keeping voltage profiles within the permitted ranges, in case of failures or transient phenomena the voltage is stabilized through a rapid supply of reactive power with which the entire system is stabilized and does not allow a general failure. Currently commercially there are several SVC models, used to solve load flow problems including PV node and PQ models of variable susceptance based on system voltage another higher precision model uses a variable susceptance which is controlled by the firing angle of the thyristors improving the range of the nodal voltage value according to the values of the selected regulations [4].

Journal of Physics: Conference Series **2065** (2021) 012023 doi:10.1088/1742-6596/2065/1/012023

One of the devices to reduce the inconveniences due to power quality issues are SVCs mainly in metallurgy industries, in the reference [5] the authors propose the use of a fuzzy logic controller to improve the voltage profile by reducing the capacitance of the device, the proposed method manages to reduce the harmonic currents with which it can compensate the effect of power oscillations, improving the quality of the energy, the novel method has a better performance compared to traditional mathematical methods.
Chan et al. [5] present the simulation of a hybrid model of power systems with a static VAR compensator, where dynamic phasor theory is applied, transient models are used for the AC subsystem, researchers first derive the model and use a connection algorithm between the phasor model and the subsystem. The model is tested on 39 buses located in New England, resulting in high model accuracy compared to the transient electromagnetic model, allowing the analysis of the dynamic response of electrical power systems containing SVC.
Zhigun et al. [6] using a new method they perform the simulation of an electrical power system with SVC where they apply the theory of single-phase dynamic phasors, transient models are used for the AC subsystem. An interface algorithm is used to join the dynamic phasor model of the SVC and the AC subsystem, the simulation results show that var's static compensator single-phase dynamic phasor model has greater accuracy than the transient electromagnetic model, with the simulation of the new hybrid method achieving a new vision for dynamic simulation of large power systems with SVC.
In this work, a power system connected to a var static compensation phasor model is simulated to regulate system voltage by generating reactive or capacitive power according to the voltage threshold to be controlled and analysing the system's response for instability or electrical failures, the results of the investigation are described in the document [7].

2. Dynamic phasor model of the SVC

2.1 Model concept

The SVC connected in the electrical power systems operate equivalent to a capacitor, by means of which they can control the voltage and reactive power in the bars or terminals near the connection, in an ideal state without loss of active power (w) or reactive (var) provided instantaneous responses to the system, manages to maintain the constant voltage, among the goodness of the device is the unlimited ability to absorb or generate reactive power [7].
The phasor dynamic model is also the lame method of averaging in the state space, it is developed based on the theorem of the Fourier series with variations in time coefficients. It is assumed that the waveform in the time domain can be represented $x(t)$ in a time interval $t =\in (t-T, t)$ applied in the Fourier series and can be calculated using the following equation:

$$x(t) = \sum_{k=-\infty}^{\infty} X_K(t) e^{jk\omega_s t} \tag{1}$$

$$\omega_s = 2\pi/T$$

Where, $\omega_s = 2\pi/T$, and is the dynamic time $X_K(t)$ phasor along with the variable coefficients in time of the Fourier series, the dynamic k-th phasor at time t is calculated using equation 2.

$$X_k(t) = \frac{c}{T}\int_{t-T}^{t} X(t) e^{-jk\omega_s t}\, dt = (X)_K(t) \tag{2}$$

Where, c s 1 for k s 0 and c s 2 for k> 0.

Two properties are used for the development of the SVC dynamic model. The first property refers to thel-difference of the dynamic k-th and is described by the following equation.

$$\frac{dX_K}{dt}(t) = \left(\frac{dx}{dt}\right)_k (t) - jk\omega_s x_k(t) \tag{3}$$

Property 2 is the product of two dynamic phasors, where the dynamic k-th phasor of the product of two-time domains with the waveform in x(t) and y(t), can be calculated using equation 4.

$$(xy_k) = \sum_{i=\infty}^{\infty} (x)_{k-1}(y)_i \tag{4}$$

The waveform in the time domain x(t) can be calculated from dynamic phasors using the following equation:

$$X(t) = Re\left(X_k(t)e^{jk\omega_0 t}\right) = X_{-K}(t)e^{-jk\omega_0 t} + X_{-(K-1)}(t)e^{-j(k-1)\omega_0 t} + \cdots$$
$$X_{K-1}(t)e^{j(k-1)\omega_0 t} + X_k(t)e^{jk\omega_0 t} \tag{5}$$

x(t) is a real function according to a time-time int, it is easy to conclude$t =\in (t - T, t)$ that, $X_{-K} = X_K^*$where * indicates that you are working with a complex variable.

2.2 TCR thyristor-controlled reactor

The advances presented in power electronics have allowed the development of thyristor-based SVC devices, the thyristor-controlled reactor (TCR) is one of the characteristic elements, the element combining fixed capacitors or switched by thyristors, whose main mission is to provide a fast and continuous control of the reactive power consumed by a coil.

The diagram of a TCR is made up of an inductive resistive coil together with an alternating current to alternating current converter formed by 2 antiparallel placed thyristors, the control is carried out by phase, where the value of the current through the inductance is regulated, the current that circulates is a transient sequence on the whole of the coil RL. For the analysis of the behaviour of thyristor-controlled reactor thyristors behave in an ideal way, the pure sinusoidal potential differential can be calculated by means of equation 6.

$$e(t) = \sqrt{2}Esen\omega t \tag{6}$$

2.3 TSR thyristor-switched reactor

TSRs are defined as a thyristor-switched inductor, effective reactance is presented in a staggered way due to complete thyristor conduction or zero conduction management if the thyristor valve is open. Thyristor-switched reactors are bypass compensators, which can produce or generate reactive power. TSRs have a simple principle of operation mentor, consist of a half cycle delay and have the peculiarity of non-harmonics [8].

The thyristors are connected in parallel; they are built by different inductors and they are switched to the operating state or through the switching of thyristors without controlling the firing angle, achieving lower costs and losses, however it does not have continuous control.

2.4 TCR Dynamic Phasor Model

The single-phase TCR circuit, the thyristors are connected in parallel but inverted with the aim of firing symmetrically in positive and negative with respect to the semi cycle of the voltage, in this connection only the current of harmonics of odd order occurs.

To prevent or eliminate the circulation of triple-order harmonics the delta connection of three single-phase TCR is used, research conducted shows that harmonics of more than the seventh order have fewer effects on dynamic characteristics of the TCR.
In many cases considers the fundamental and fifth harmonic for the dynamic phasor model of the TCR [9], for the calculation of voltage and current of the single-phase circuit the following equation is used.

$$\begin{cases} C\dfrac{dy}{dx} = i_l - i \\ L\dfrac{di}{dt} = SV \end{cases} \tag{7}$$

where, the variable S is there presentation of as witching function defined by S=1, when one of the two thyristors is energized and driving, when S =0 the thyristors are in an open state.

2.5 *Dynamic phasor model of filter circuits*

The RLC filters are made up of resistors, inductances and capacitors connected in series, these elements are connected to the SVC bus to eliminate high-order harmonics. Applying the differential property in equation 8 is derived in e circuit resistive, inductive, separates the real and imaginary parts in equation 8 and 9, infers the model of the circuit is the $V_1(t)$voltage, the same procedure is applied for the current.

$$V_1(t) = L\frac{di(t)}{dt} + Ri(t) \tag{8}$$

the dynamic phasors are calculated:

$$L\frac{dI_k}{dt} = V_{1,k} - jk\omega_s lI_k - RI_k \tag{9}$$

2.6 *Var static compensator block*

Simulink in the library presents the block the phasor model of the three-phase static compensator of var, in the connection terminals regulates the voltage by controlling the amount of reactive power generated (capacitive SVC) when the voltage is low and absorbs reactive power (inductive SVC) when the voltage is high, Switching three-phase capacitor banks and secondary inductors installed in a coupling transformer varies the reactive power of the device.
The compensators are turned on and off by means of three thyristor switches (TSC), while thyristor-controlled (TSR) or phase-controlled (TCR) reactors are turned on and off. Figure 1 shows the one-line diagram of the SVC next to the simplified block diagram of the control system [10].

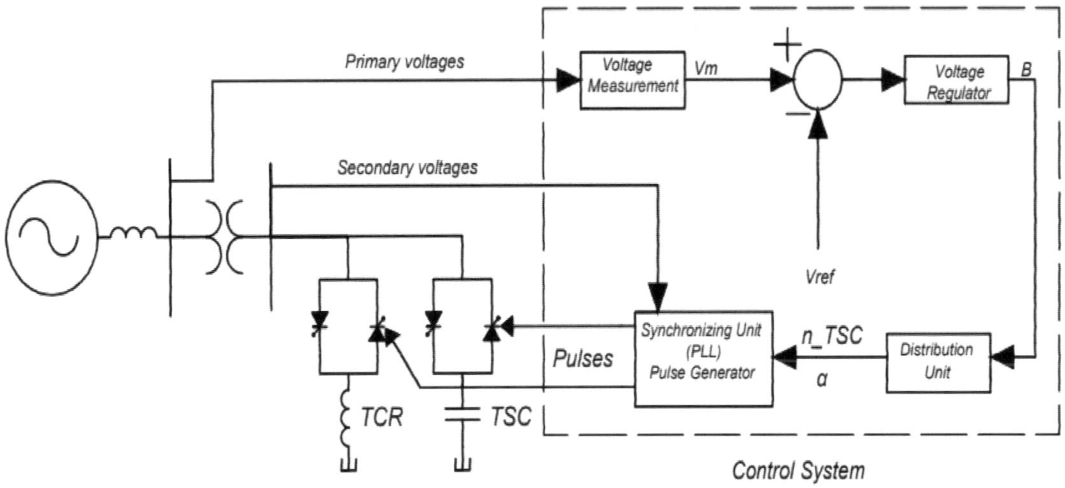

Figure 1. Diagram of a line of an SVC and its control system block

In the control system the positive sequence voltage to be controlled is measured, the measurement system is based on Fourier, using the moving average of a cycle, the voltage regulator uses the difference known as error between the measured voltage and the reference voltage to calculate the B susceptance of the compensator to maintain the constant voltage.
The SVC block (phasor type) is a phasor model, which has to be used with the phasor simulation method, it includes elements such as synchronous, various dynamic loads to carry out the analysis and study of transient stability and observe the impact of the static compensator on the power system.
In control parameters the block is configured in voltage regulation mode, it is necessary to parameterize the reference voltage in pu, the signals are processed in the signal processing block, where the characteristics of the sampled information are extracted, analysed in a uniform way and not uniform.
The mentioned block can contain filters for resampling and smoothing to extract the characteristics of change point, patterns etc. Additionally, it can analyse several signals simultaneously in time and frequency domains.

3. Power system modelling

The system is modelled in Matlab Simulink, the goal is to analyse the response of the behaviour of the SVC against instability and network failures, the SEP is connected to a voltage of 500 kV, 60 hz, when the system voltage is under the compensator generates capacitive reactive power by means of the TSC (thyristor controlled capacitors) with a capacity of 47.1 MVar.
While when the voltage is high it absorbs the inductive reactive power by means of TCR (thyristor-controller inductors) with a capacity of 97.6 MVar, the SVC block used is of phasor type, pre-work is referenced [11] electrical failures the three phase fault block is used, the opening and closing times can be controlled from an internal timer, additional can be scheduled phase-by-phase failures, phase-to-ground failures, the block is commanded by the Three -Phase Breaker block. The figure 2 shows the plant used for simulation, with each of the blocks used to achieve the simulation of the power system.

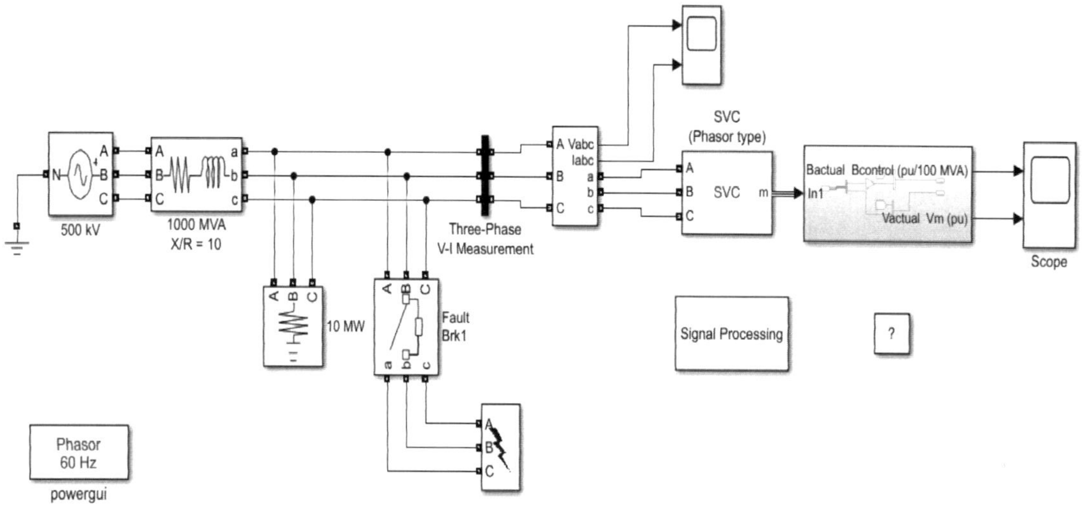

Figure 2. Block diagram used in Simulink

3.1. *Response from the SVC system*

The first simulation is to analyse the dynamic response of the SVC versus instability is in the network, for which the three-phase voltage source is programmed, at the beginning the source generates the rated voltage, after t=0.1s the voltage is reduced to 0.97 pu, after t=0.4s the voltage increases to 1.03 pu, at the end of the simulation at t=0.7s the voltage returns to a nominal value.

Figure 3 shows the dynamic response of the system, the black curve represents the positive sequence voltage of the actual system, while the blue curve is the voltage measurement with SVC response.

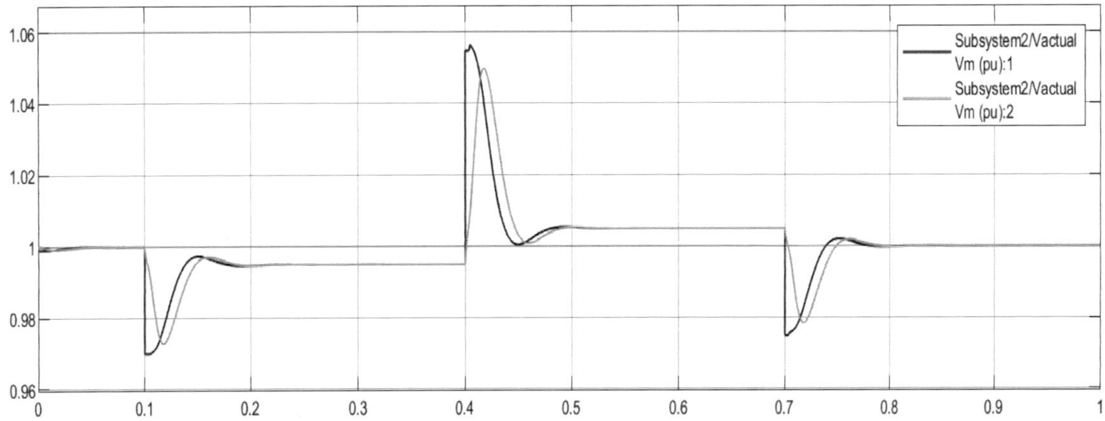

Figure 3. Dynamic response of the SVC to network instability

The actual voltage drop is 0.03 pu at that time the current of the device is fully capacitive for such a reason from t=0.2 to t=0.4 the voltage approaches 1 pu, while when the voltage increases to 1.03 pu by t=0.4 the current is inductive in order to reduce the actual voltage of the system.
On the other hand, the figure 4 shows the behavior of the curve, the voltage drop reaches 0.69 pu, however despite the performance of the SVC is not able to maintain stable voltage.
The following analysis corresponds to an electrical fault in one of the ground phases, to perform the simulation at the voltage source the types of variation are canceled either by amplitude, phase or frequency, the procedure is to close the fault breaker and program the electrical fault for a switching time(s) [20/60 30/60].

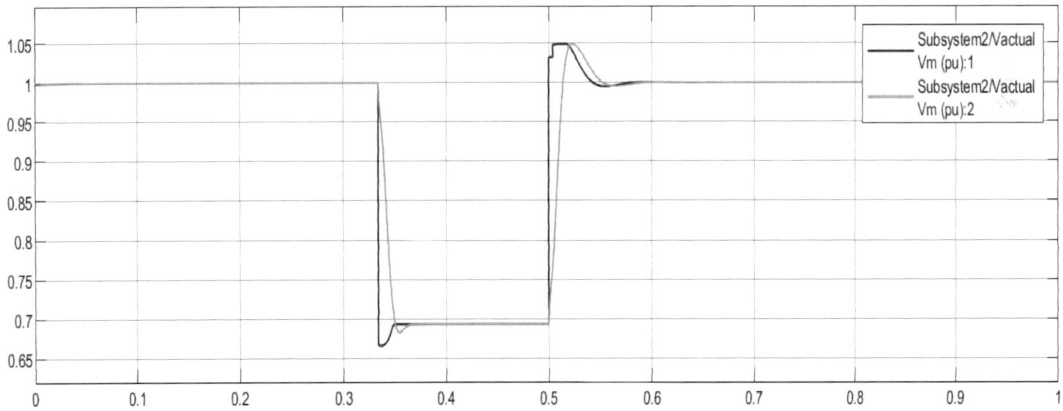

Figure 4. Dynamic SVC response to phase-to-ground failure

On the other hand when a phase-phase failure occurs the voltage drop reaches 0.52 pu, similar to the previous case you have an SVC response but due to the type of failure is not able to maintain the voltage, without clutch when you have the transient fault clearing in t=0.5 the voltage damping is displayed due to the performance of the inductive capacity of the SVC. The system response is shown in figure 5.

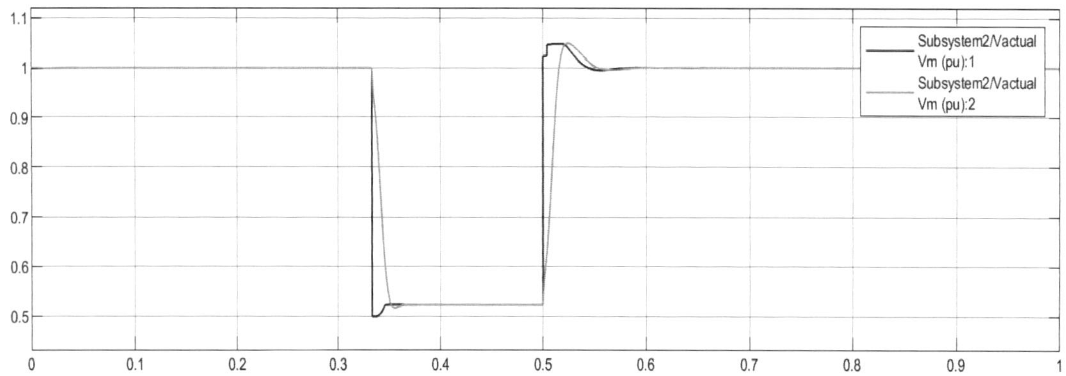

Figure 5. Dynamic SVC response to phase-phase failure

4. Conclusions

One of the devices used to control the voltage level in the SEP is the SVC, in this document an electrical system is simulated for three scenarios in Simulink, despite the difficulty of presenting the exact conditions of the system, in the first scenario presents an unbalance of the voltage level as well as drop and overvoltage, the device manages to mitigate the problem with which the network is stabilized with an acceptable response reaching up to 1.05 pu.

The second scenario is due to a phase-earth fault, in which the static compensation device is not capable of solving the problem with a voltage drop up to 0.69 pu, but it does collaborate to stabilize the problem in less time when the fault is cleared. , in the third scenario with a phase-phase failure is similar to the second, the only difference is in the depth of the voltage drop with a drop of 0.52 pu. In the following research works, other factors must be included and different values of the integral gain of the SVC must be placed to improve the dynamic response.

5. References

[1] AL Ahmad A, Sirjani R (2019) Optimal placement and sizing of multi-type FACTS devices in power systems using metaheuristic optimisation techniques: An updated review. Ain Shams Eng J 11:611–628

[2] Jordehi AR (2015) Particle swarm optimisation (PSO) for allocation of FACTS devices in electric transmission systems: A review. Renew Sustain Energy Rev 52:1260–1267

[3] Muhammad Y, Khan R, Raja MAZ, Ullah F, Chaudhary NI, He Y (2020) Solution of optimal reactive power dispatch with FACTS devices: A survey. Energy Reports 6:2211–2229

[4] Estudio del efecto del Compensador Estático de Reactivos en el problema de Flujo de Carga. http://scielo.sld.cu/scielo.php?script=sci_arttext&pid=S1815-59012012000100001. Accessed 14 Mar 2021

[5] Naderipour A, Abdul-Malek Z, Heidari Gandoman F, Nowdeh SA, Shiran MA, Hadidian Moghaddam MJ, Davoodkhani IF (2020) Optimal designing of static var compensator to improve voltage profile of power system using fuzzy logic control. Energy 192:116665

[6] Zhijun E, Chan KW, Fang DZ (2007) Hybrid simulation of power systems with dynamic phasor SVC transient model. In: Proc. Int. Conf. Power Electron. Drive Syst. pp 1670–1675

[7] Zhijun E, Fang DZ, Chan KW, Yuan SQ (2009) Hybrid simulation of power systems with SVC dynamic phasor model. Int J Electr Power Energy Syst 31:175–180

[8] 8. Goli P, Shireen W (2014) PV powered smart charging station for PHEVs. Renew Energy 66:280–287

[9] Igbinovia FO, Fandi G, Svec J, Muller Z, Tlustý J (2015) Comparative review of reactive power compensation technologies. In: Proc. 2015 16th Int. Sci. Conf. Electr. Power Eng. EPE 2015. Institute of Electrical and Electronics Engineers Inc., pp 2–7

[10] Mathur RM, El-Hawary ME THYRISTOR-BASED FACTS CONTROLLERS FOR ELECTRICAL TRANSMISSION SYSTEMS.

Journal of Physics: Conference Series **2065** (2021) 012023 doi:10.1088/1742-6596/2065/1/012023

[11] SVC (Phasor Model) - MATLAB & Simulink - MathWorks América Latina. https://la.mathworks.com/help/physmod/sps/ug/svc-phasor-model.html. Accessed 24 Apr 2021

AUTHOR INDEX